What Executives Get Wrong About SEO

The Hidden Reason Companies Fail

The SEO Trilogy - Book I

What Executives Get Wrong About SEO

The Hidden Reason Companies Fail

By Cristobal Varela

ISBN: *978-1-971029-07-8*
Publisher:
VarelaPublisher.com
Mesa, Arizona, USA

First Edition

Book I of the SEO Trilogy

Printed in the United States of America

Disclaimer:
This book is based on the author's professional experience and is intended for informational purposes only. It does not guarantee specific results and should not be interpreted as financial or legal advice.

Design and layout: *Cristobal Varela*
Cover design: *Cristobal Varela*

Also available in Spanish as
Lo que los ejecutivos no entienden sobre el SEO
Visit: https://TrilogiaSEO.com

Contents

DEDICATION

To my wife, Sonia Varela (mi persona favorita), for an amazing twenty-two years of sharing our lives. Thank you for always being there and supporting me. Your patience, love, and enthusiasm for chasing new goals as a family have been the best part of this journey.

To Santiago, my eldest son, for the courage to pursue Nuclear Engineering in the U.S. Navy.

To Paulina, my beautiful princess—the creative soul of our home—whose strong character drives her to pursue Genetic Engineering.

To my always-smiling Julián, who has helped us understand that our knowledge is limitless.

To my beloved parents, Jorge A. Varela and Betty L. Varela, for the immense sacrifice of coming to a country where everything was different so my siblings and I could have a better future. Thank you, from the bottom of my heart.

To my siblings—Omar, Vianett, and Danyra—for always standing together. The memories, joy, laughter, shared successes, and support in times of need have only made us stronger. I love you.

ACKNOWLEDGMENTS

To my Sensei, David Lopez, from Ki-Senshi Martial Arts, thank you for sharing not only your achievements and ideas, but also the discipline and perspective that extend far beyond the dojo. You were the one who started me on this path of publishing a book and reminded me of that simple life goal: to plant a tree, raise a child, and write a book. This project exists, in many ways, because you helped me see that last part as possible and necessary.

To all my clients from 1998 to 2025—thank you for your trust and for the opportunity to grow together across so many industries. The lessons we learned side by side are distilled here to help other leaders make better decisions.

To the hiring managers who opened the door, and the leaders who backed my SEO strategies and pushed for real implementation—thank you for believing in the work and giving it room to prove its value.

Special thanks to my support team— Juan Pablo Hudec, César Gutiérrez, Luz Tania Hernández, Jorge Varela, Josué Álvarez, Ernesto López, Melanie Rocha, Guillermo López, Marcos Carrasco, Nadia Vigil, Mónica Sánchez, Franco Escobar, and Gabriel Castellanos. Your skill and commitment make this work possible.

To Jesus Segura, photographer and family friend, who generously offered to take my portrait without hesitation. Thank you for your kindness.

AUTHOR'S NOTE

The stories in this book are based on real experiences I have had working with companies of many sizes, across multiple industries, over many years. The decisions, challenges, and outcomes described here are true to the best of my knowledge.

To respect the privacy of former coworkers, supervisors, and clients, some names, job titles, and identifying details have been changed. In a few cases, I've also adjusted timelines or combined similar situations from different companies into a single narrative.

None of these adjustments change the core lessons. The purpose of this book is not to expose individuals, but to help leaders recognize patterns that either unlock or undermine the full potential of search engine optimization. My hope is that you will see your own organization more clearly—and use these insights to make better decisions for your teams and your business.

PREFACE

My first real lesson in SEO did not come from a big brand or a complex enterprise site. It came from a small company called Van Hees Stoneworks, LLC.

I built them a simple website, thinking my job was mostly design and basic setup. When they started appearing in the first position on Google, I realized something important: the website was only half the work. What really mattered was whether it produced a return on their investment.

One afternoon, Werner, the owner of Van Hees Stoneworks, walked into my office with a check for $2,500 USD. He put it in my hand and said, "Keep doing whatever you are doing with the website. I closed three deals today just because of your skills." That was the moment it clicked for me: SEO was not about traffic for its own sake. It was about revenue, pipeline, and real outcomes for real businesses.

A second moment drove the point even deeper.

An executive named Peter, the CEO of a company called GMSI, called me one day and said, "Cristobal, it's urgent that we have a meeting." His tone immediately made me think something was wrong. As he spoke, I was already checking his website, expecting it to be down or broken. He gave me no details over the phone, just, "Please come to the office."

When I arrived, his door was half open. I could see him staring at it, waiting. As I stepped in, his business partner, who had been standing behind the door, jumped in front of me and said something like, "Tell me to jump, and I'll ask how high." I was confused and a little nervous.

Peter stayed serious, looked at me, and said, "This is what happened," as he slammed a document onto the desk. It was a paid invoice. The only thing that really registered at first was the number: a seven-figure sum. More than one million dollars in revenue from a contract with a Taiwanese company—won, in large part, because of the structure, clarity, and visibility created by SEO.

These two companies started referring more clients than I could realistically take on. I had to turn many of them down. The ones I did accept came in expecting the same kind of results in a week.

That is when I began to see the real disconnect: business owners, marketing managers, and other leaders wanted the outcomes of mature SEO, but they expected them on a paid-media timeline. They wanted million-dollar contracts with the patience of a seven-day campaign.

From that point on, I started taking notes. I documented what worked, what did not, and where people were tempted to cut corners. I studied the difference between white-hat practices that build durable value and black-hat shortcuts that create short-lived spikes and long-term risk. It became obvious that "quick fixes" were not feasible if you cared about the health of the business.

What separated the success stories from the disappointments was not just tactics. It was maturity—on the part of leadership. The leaders who understood that SEO is a long game were the ones who ended up with the strongest, most defensible growth.

Why I wrote this book

When I look back at almost every company I have worked with, there is a pattern that still bothers me.

As a person, I was appreciated. People were kind. I was paid well. But when it came to SEO, almost no one listened—and the few who did had other "urgent priorities." I watched companies spend thousands, sometimes millions, on campaigns that made noise but did not bring a real return on investment. Every time, it was the same story: paid media first, redesigns first, "quick wins" first. SEO last.

There was not a single one of those jobs where I did not think seriously about quitting. But then I would ask myself: if I walk away, we both lose. They lose the chance to build something sustainable, and I lose the chance to prove what SEO can really do. And in that scenario, who gave up? I did not want the answer to be me. The question became: how do you help leaders understand something they keep pushing to the bottom of the list? You cannot force them. You cannot just go around your manager to the Vice President of Marketing every time, either.

So I chose a different path: document everything.

I wrote recommendations to improve indexation and visibility. I prepared step-by-step documents explaining what to do, why it mattered, what we expected to see, and how it aligned with their business goals. Anything that could help catch leadership's attention, I tried. If they rejected the proposal, I still did my job: designing strategy and handing it off. Over and over again.

Until one day, a leader called me into her office and said, almost accusingly, "SEO is not working. We hired you to do this right. Are you not being paid well enough?" I smiled and answered honestly: "Everyone here is very kind, and yes, I'm paid well. But SEO is the last priority on the list."

"How can that be?" she asked.

I showed her my recommendations—what needed to be done and when I had delivered them. "I was told there was no time to implement any of this," I said. "Who told you that?" she asked. "Brendan."

She called him in. "Why have we not implemented these SEO recommendations?" His answer was simple: "Lorie, you asked me to work on new site features and said nothing was more important than that. So I did not implement SEO."

That same dynamic repeated itself almost everywhere I worked.

I am writing this book for both sides of that conversation.

I want to help leaders truly understand what SEO is, how it works, and why it cannot keep living at the bottom of the priority list if you expect meaningful results. And I am writing it for all the SEO professionals who quietly spend their days building strategies, checking technical details, and crafting content that is designed to convert—only to have their work ignored, delayed, or dismissed as "nice to have."

We deserve a seat at the table—not for ego, but because technical details and strategic decisions are deeply connected. Someone needs to be able to explain, clearly and confidently, why a JavaScript function is not the best way to render critical content, why "noindex" must be removed from live production sites and reserved for staging, and why the small local company holding the first position on Google is often a more serious threat than the enterprise brand that does not even appear in the top 100 results.

SEO professionals can help your site perform better, your marketing budget work harder, and your brand compete where it really matters. We are not, and never have been, the last item on the list.

This book exists so leaders and SEO practitioners can finally meet in the middle—with shared language, mutual respect, and decisions that create long-term value instead of short-term illusions.

INTRODUCTION

• The cost of misunderstanding SEO

Search engine optimization isn't a buzzword, a hack, or a task for the intern. It's a long-term investment in your company's visibility, reputation, and revenue. But for most companies, SEO is either misunderstood, delegated blindly, or ignored until it's too late.

You don't need to be an expert in SEO strategies, but you do need to understand what it is, how it works, and what role it plays in your business. If you're signing off on marketing budgets or approving vendors, your understanding of SEO can either fuel growth—or silently sabotage it.

Over the years, I've seen organizations that have spent tens of thousands of dollars on SEO with nothing to show for it. Not because SEO doesn't work—but because leadership didn't know what to ask, what to measure, or what to expect. That misunderstanding creates blind spots, and those blind spots cost companies more than they realize.

• The myth of the quick fix

Many leaders still see SEO as a switch to flip or a problem to "fix." They notice rankings drop, conversions slow, or visibility shrink—and assume something is broken that can be repaired with a plugin, a keyword, or a quick update. But SEO doesn't work like that.

Search visibility depends on **hundreds of interconnected variables**: site structure, load time, usability, content depth, schema markup, and trust signals across the web. It's not one dial—it's a soundboard with many knobs that need to be tuned together.

That's why the best SEO outcomes don't come from one tool or one campaign. They come from alignment—between leadership expectations, marketing execution, and technical readiness. When one of those is missing, no "fix" will last.

The goal of this book isn't to simplify SEO into a checklist. It's to help you see the *system* behind it—the coordination that separates companies that grow steadily from those that chase temporary wins.

• Who this book is for (and not for)

This book is for business owners, marketing leaders, startup founders, and executives who want to understand SEO at a strategic level. You know marketing. You don't need to know code. What you need is a clear understanding of how SEO works so you can make better decisions, hire the right people, and protect your investment.

If you're a hands-on SEO specialist looking for advanced technical tactics or link building schemes, this isn't the book for you. But if you're tired of hearing conflicting advice and you want a reliable, strategic overview of how SEO actually impacts your business, you're in the right place.

• What this book will teach you

This book will not teach you how to optimize a page or run a crawl audit. Instead, it will show you:

- What SEO really includes (and what it doesn't)

- Why so many SEO efforts fail at the leadership level

- How to evaluate SEO talent, agencies, and proposals

- How to align SEO with your growth, branding, and conversion strategies

- How to avoid the most common mistakes that cause wasted time, money, and visibility

You'll also learn what search engines are evolving into—how AI, language models, and zero-click results are changing the game, and what that means for your digital strategy.

• How to use this book

You can read this book front to back, or jump into chapters based on your current challenges—hiring, content, performance, or reporting. Each section is designed to be self-contained and practical. No chapter assumes you've read the one before it.

If you're new to SEO, start at the beginning. If you've managed SEO before but want clarity and a fresh perspective, use the chapter titles as a menu. This book is here to give you the confidence and context you need to lead smarter conversations, make better decisions, and avoid common traps.

We've also included a Content Scorecard later in the book—a framework you can use to evaluate pages or campaigns at a glance. It's not a tool you download. It's a system you understand and apply, because knowing *what good SEO looks like* is half the battle.

• Why this book needed to be written

There are plenty of SEO books out there—most of them written for practitioners. This one is written for leaders.

Most executive teams still treat SEO like a checkbox or an afterthought. It's usually handed off, underfunded, or misunderstood. Meanwhile, the companies that win at SEO have one thing in common: leadership that gets it.

This book is meant to help you get it—not by becoming an expert, but by becoming fluent in the decisions that matter. That fluency will protect your company from wasted budget, bad hires, misleading reports, and shallow tactics.

If you understand what SEO really is—and what it demands—you won't just approve SEO. You'll lead it the way it needs to be led.

• Understanding what SEO really means

Search Engine Optimization (SEO) is not about algorithms—it's about people. At its core, SEO is **user experience at its best**. Google's ultimate goal is to reward websites that deliver trust, clarity, and satisfaction to the person searching. Every click, scroll, and interaction sends a signal about whether a site fulfills its promise.

When users find what they're looking for—quickly, confidently, and without frustration—Google sees it as a sign of quality. That's why the most effective SEO strategies start with improving the experience, not chasing rankings.

Beneath that philosophy, SEO connects three essential layers:

- **Technical foundation** — A fast, secure, and well-structured site that loads smoothly and lets search engines understand its content.

- **Content relevance** — Information written for humans first, answering real questions with depth, accuracy, and intent.

- **Authority and trust** — Validation through quality links, mentions, reviews, and engagement that signal reliability.

When these three elements align under a great user experience, SEO becomes a system that grows on its own. Every optimized page, every helpful article, and every satisfied visitor strengthens the signals Google values most.

Unlike paid ads that disappear when the budget ends, SEO compounds. It builds digital trust over time, turning your website into an asset that attracts new customers day after day.

In simple terms, SEO is the bridge between user satisfaction and discoverability that drives business results—leading naturally to Search Experience Optimization (SXO), a smarter way to think about SEO.

PART I: AI and the new SEO landscape

1. SEO is not dead, it has diversified.

Since the public release of **ChatGPT in November 2022**, the digital landscape hasn't stopped shifting. And every few months since, someone new declares SEO dead.

They point to ChatGPT's rise, falling click-through rates, or Google's evolving results pages—and assume the game is over. But it's not. SEO isn't dead. In fact, it's more critical than ever. It just **evolved faster than most companies could adapt**.

What's dead is the old version of SEO—the one that treated optimization as a simple checklist: add keywords, get backlinks, repeat. That approach used to work. It doesn't anymore.

Today, SEO is a **multi-layered system** that blends technical precision, content clarity, user experience, and machine comprehension. And at the center of this shift is artificial intelligence (AI).

The rise of **AI engines** has redefined what it means to "rank." Instead of listing websites, these systems synthesize answers—surfacing results based on content quality, structure, and trust. And the sources they pull from? They're not random.

Tool tip: *AI engines.*
Platforms like Search Generative Experience (SGE), Bing Copilot, and ChatGPT that use artificial intelligence to process queries, generate responses, and sometimes assist with tasks. They don't always behave like traditional search engines—some are conversational, others are exploratory, and not all cite sources.

They prioritize **structured, authoritative, and machine-readable websites**. Content that's not only helpful to humans, but also understandable to machines.

We've officially moved from **search engines that match keywords** to **AI engines that interpret meaning**. This changes everything.

These engines are powered by **Large Language Models (LLMs)** that don't just crawl web pages—they understand them. They evaluate **semantic signals**, **topic relationships**, and **schema markup** to determine not just *what your content says*, but *what it's about*.

Tool tip: *Large Language Model (LLM)*
A type of artificial intelligence trained on massive text datasets to understand and generate human-like language. Examples include GPT-5, Claude, and Gemini.

So no, SEO isn't dead. It diversified—into something more sophisticated, more interconnected, and more demanding.

2. Invisible SEO that moves revenue

In July 2025, leadership asked me what our strategy was for AI visibility.

My response: *"That was implemented last year."*

Back in 2024, I led a sitewide overhaul of structured data. It wasn't flashy. There was no big announcement. Just precise **schema markup** built around the company's actual business model—communities, homes, floor plans, location hierarchies, and pricing structures.

It wasn't called an "AI strategy" at the time. But that's exactly what it was.

Tool tip: *Schema markup*
A standardized way of tagging content so that search engines and AI engines can understand the meaning and structure of a page—not just its text.

When their analyst pulled the data, the outcome surprised everyone—except me.

Organic visibility across AI engines grew by 3,989% in 12 months.
No paid media. No extra content. Just structure.
Just clarity—for machines and users.

And yet, in those early months, I faced the usual resistance:

- "Why do we need this if the site looks good?"

- "Will this make our site prettier?"

- "Is Google even using this data?"

The truth is, most leadership teams and web developers *don't* fully understand SEO strategy when it's explained—only when it shows up in results. And by then, it's too late to have been proactive.

That's why this book exists.

3. What executives need to know about AI

In the age of artificial intelligence, your company's visibility no longer depends solely on Google rankings. AI engines are answering questions directly, summarizing sources, and increasingly acting as the first point of contact between users and information. For your content to be included in that process, it must be structured and trusted at a machine level.

This is where **GEO**, **AIO**, and **AEO** come in. These are not trends. They are the framework for digital visibility moving forward.

- **GEO – Generative Engine Optimization**

Focuses on **preparing your content to be cited, quoted, or summarized by AI engines** like ChatGPT, Perplexity, and Google SGE. This means aligning your structure, tone, and topical clarity to systems that synthesize—not just index.

- **AIO – AI Optimization**

Involves adapting your entire digital footprint (not just individual pages) to be legible to large language models. This goes beyond SEO—it touches product pages, customer support content, even internal documentation that AI systems may learn from.

- **AEO – Answer Engine Optimization**

Targets structured, authoritative content that AI engines can extract as direct answers. This often includes clear **H-tag hierarchy**, question-and-answer formatting, authorship signals, citations, and schema markup designed for clarity and trust.

Tool tip: *GEO/AIO/AEO*
Emerging frameworks designed to increase your brand's inclusion and performance within AI-powered environments. These approaches prioritize clarity, structure, and credibility for machines that summarize content rather than list links.

You don't need to lead implementation. But as an executive, you need to know if your team is addressing these concepts—because competitors who are will begin to replace you in ways that are difficult to track... until it's too late.

Note: As critical as this AI-driven visibility is, it's only half the equation. Once users find you, they still need to take action. That's where **Conversion Rate Optimization (CRO)** becomes essential. It requires a different level of expertise—focused on turning attention into outcomes—and we'll address that fully later in this book.

Why AI engines matters for your business

Many companies haven't noticed the shift because they're still measuring traffic the old way. They're looking for visits and pageviews. But modern discovery is happening in systems where no one ever clicks a link.

Visibility now means being:

- Summarized accurately by AI

- Cited by machine-generated answers

- Indexed deeply enough to be found through structured queries

- Trusted by engines that evaluate not just what you say, but how you say it

If your website isn't built with this in mind, you're invisible—even if you rank.

And the worst part? You won't know it until you've already been replaced.

Takeaways: The Search Shift Already Happened

- **Problem:** AI engines now synthesize answers without sending users to your website.
  Fix: Make your content machine-readable using clear structure and schema markup.

- **Problem:** Traditional SEO tactics (just keywords and backlinks) no longer move the needle.
  Fix: Align your SEO efforts with modern pillars: structure, speed, UX, authority, and data.

- **Problem:** Your company may be invisible in AI-generated answers and not even know it.
  Fix: Audit how your content appears in tools like ChatGPT, Perplexity, and Google SGE. If you're not being cited, ask why to your SEO team.

- **Problem:** SEO success often goes unrecognized because it's not tied to business goals.
  Fix: Connect SEO performance to meaningful outcomes—**lead generation, conversions, and revenue—not just traffic.** Show how organic visibility supports the customer journey, not just the click.

- **Problem:** Leadership teams are often reactive—understanding SEO only after results appear.
  Fix: Build fluency now. You don't need to do SEO, but you do need to lead it with informed questions and strategic oversight.

4. Algorithm updates are business updates

Algorithm updates are often framed as technical housekeeping—something the SEO team tracks in the background. That framing is outdated. Every significant algorithm change is a business event. It influences how your company is discovered, interpreted, and trusted. If your visibility drops, it's not merely a ranking problem. It's a growth problem.

These aren't minor changes. They reset the field.

Google releases thousands of adjustments each year, most small. Several times a year, it releases broad re-weightings called **Goolge core updates** that can move leaders to the middle and elevate previously overlooked contenders.

Do not panic: stability beats scramble

In my experience, the worst reaction to a core update is a hasty attempt to change everything. Most of the time, **a solid structure holds**. If your site is technically sound—clear architecture, crawlability, clean internal linking, schema, authorship signals, and useful content—traffic often **recovers within a few weeks to a month** as Google reassesses quality across domains.

One case stands out: a news site lost half its organic traffic after a core update. Leadership was understandably alarmed. However, sister sites on the same network, running the same CMS and built with identical structural principles, were unaffected. That difference gave us enough confidence not to dismantle anything prematurely. We observed   carefully, stayed the course, and within a month, the site regained visibility. The foundation worked.

Not every organization has the luxury of sibling properties for comparison. But the principle still holds: **make no structural changes until you've evaluated the full context**. Reacting too soon can compound the problem.

Not all algorithm updates are the same

- **Core updates:** Broad recalibrations that reweight value signals across industries and query types. Rankings can shift even when nothing on your site is "broken." Sites with clear intent match, depth, strong internal linking, fast pages, and credible authorship tend to gain; thin, unfocused, or slow experiences tend to slip.

- **Spam updates:** Target manipulative tactics such as link schemes, doorway pages, and spun/autogenerated content. Impact is often sharp (demotions or deindexing) and persists until the abuse is removed. The path forward is remediation (remove/replace, noindex where needed), cautious link cleanup, and rebuilding with genuine signals.

- **Helpful content updates:** Downrank pages created to chase rankings rather than serve people. Originality, first-hand expertise, clear purpose, and satisfying answers are rewarded; fluff, clickbait layouts, and latency are punished indirectly through poorer engagement signals. Audit intent alignment and usefulness first, then refine structure and performance.

How to respond to algorithm updates—without overreacting

Executives don't need to predict every algorithm change. But when rankings drop, they do need a clear framework for decision-making. The wrong reaction—rushing to change content or structure—can do more harm than good. The right one begins with **observation**, followed by **pattern recognition**, then **action**.

Here's how to lead with confidence:

When to hold steady:

- The drop occurs right after a confirmed **core update**

- Your site has strong fundamentals (speed, structure, E-E-A-T, schema)

- Your industry peers or "sister sites" are also impacted

- Traffic stabilizes or improves within 2–4 weeks

When to initiate a response:

- The decline is isolated and no update has been confirmed

- The drop continues **beyond 30 days** with no recovery trend

- Your competitors are gaining traction while you're losing visibility

- You identify clear gaps in **E-E-A-T, technical SEO**, or **content clarity**

Once you've assessed the situation, here's how to act:

- **Reinforce your SEO foundation first.**
  Confirm indexability, crawlability, content quality, and internal linking before adjusting strategies.

- **Avoid reflexive overhauls.**
  Recovery often takes time—premature changes can remove what's actually working.

- **Harden E-E-A-T where it matters most.**
  In law, health, finance, or media—author bios and credentials aren't optional.

- **Unify your teams.**
  Treat technical SEO and content not as departments—but as a system.

- **Expect resistance. Lead anyway.**
  Updating bios or author pages seems trivial. It isn't. Every trust signal counts in recovery.

- **Track performance beyond rankings.**
  Focus on qualified leads, conversions, and visibility across AI-powered platforms.

Takeaways: Navigating Algorithm Updates

- **Problem:** Ranking volatility causes panic and overreaction.
  Next step: Don't dismantle your site. Let it stabilize before making structural changes.

- **Problem:** You have no benchmark for how the update affected your category.
  Next step: Compare across internal patterns or documented industry reports. If you have sister sites, use them to inform—but not dictate—your response.

- **Problem:** You're in a sensitive industry, and trust signals are weak.
  Next step: Improve author bios and transparency markers. Show credentials, not just opinions.

- **Problem:** Editorial teams push back on foundational updates.
  Next step: Frame trust work (bios, structure, schema) as strategic—not cosmetic. Show how it fits into the bigger system.

- **Problem:** Technical and content strategies are disconnected.
  Next step: Unify both sides. You can't fix rankings if Google can't access or interpret what you publish.

- **Problem:** SEO wins aren't connected to real business KPIs.
  Next step: Track visibility alongside conversions and lead volume—not just keyword positions.

- **Problem:** Your SEO team operates in isolation. **Next step:** Require post-update summaries that link performance shifts to actionable changes.

5. E-E-A-T is a strategic framework

At the heart of many of these updates is a consistent evaluative model: **E-E-A-T**—Experience, Expertise, Authoritativeness, and Trustworthiness.

E-E-A-T is a quality framework Google uses to evaluate whether content should be trusted and surfaced. It emphasizes:

- **Experience:** Has the content been created by someone with real-world familiarity?

- **Expertise:** Is the author formally knowledgeable or qualified on the topic?

- **Authoritativeness:** Is the content published by a recognized source or brand within its field?

- **Trustworthiness:** Is the information transparent, cited, current, and easy to validate?

This framework matters most for websites in industries where trust is essential—**law, healthcare, finance, and journalism**. In these sectors, Google expects real-world expertise, not marketing gloss.

One clear tactic that strengthens E-E-A-T: **improved author profiles**. Pages written by unnamed or vague contributors fail to inspire trust. But when authors are linked to bios that show credentials, years of experience, and proven authority—trust improves.

- A law firm's site should show bar memberships and practice specialties.

- A health content site should show medical credentials, clinic affiliations, or research experience.

- A news outlet should show the reporter's beat, tenure, and editorial standards.

This isn't about inflating titles. It's about helping machines—and users—understand that there's a real, qualified person behind the information.

Organizational pushback is predictable—and solvable

Even within reputable organizations, resistance is common. Editorial leaders often resist profile updates, structured metadata, or even linking to internal author pages. It's more work. And on its own, no single tactic seems worth it.

They're right. One tactic alone isn't enough.

SEO doesn't hinge on one variable. It hinges on **how signals are layered**. Updating bios won't lift a site out of decline alone—but paired with improved crawlability, strong internal linking, high-quality content, and topical clarity, it becomes a powerful part of a larger system.

Best of all? Author pages can often be built or improved in a single afternoon. It's low effort, high signal—**if you make it part of your strategic baseline.**

Why E.E.A.T. matters for your business

Google is not just evaluating content. It's evaluating the **source** of that content. In a post-AI landscape flooded with automated and unverified information, trust signals are no longer optional.

If your company operates in an industry that affects people's health, finances, safety, or major life decisions, your content will be held to a higher standard. That includes:

- Having clear authorship and attribution

- Showing evidence of experience and credibility

- Demonstrating a real-world presence (offices, certifications, contact info)

- Providing supporting signals like citations, editorial standards, or external reviews

A lack of these doesn't just affect rankings—it affects brand trust, user engagement, and referral potential from AI engines. If an AI system can't determine who you are, what authority you hold, or why your page should be cited over others, you'll be excluded from the conversation entirely.

You don't need to understand every algorithm detail. But you do need to ensure your organization is **discoverable, verifiable, and trustworthy**—especially in high-stakes topics.

6. SEO is not a department: it's a system

SEO is not a department. It's a **cross-functional discipline**—an operating system that runs quietly beneath your entire digital presence. It doesn't live in one role, one report, or one team. It touches everything: your website, your marketing, your product experience, and your brand's credibility. It's not a short-term tactic or a campaign add-on. It's a **long-term engine** for visibility, trust, and growth. When it's treated like a task—delegated, siloed, or reduced to keywords—it stalls. But when treated like a **system**, it compounds.

You hired an agency. You posted a job. You ran an audit. Maybe someone on your team reports on rankings or traffic every month. But here's the truth: **none of that means you have SEO.** You may have *touched* SEO—but unless it's part of your system, it's not part of your growth engine.

SEO Pillars: Why Executives Need a Broader Lens

Practitioners are often taught that SEO is built on three core pillars: **technical**, **on-page**, and **off-page**.

- **Technical** – This is the concrete foundation your entire website is built on. It ensures that search engines can access, understand, and evaluate your pages. That includes site speed, mobile responsiveness, **crawlability**, and proper **indexing**. When this foundation is weak, everything built on top—content, UX, branding—becomes unstable. It doesn't matter how good your messaging is if the infrastructure prevents people (and search engines) from seeing it.

Tool tip: *Crawlability and indexing.*
Crawlability refers to how easily search engines can access the pages on your website. Indexing determines whether those pages are stored and eligible to appear in search results. A page must first be crawlable before it can be indexed. If either fails—due to broken links, blocked files, or poor structure—your content remains invisible, no matter how good it is.

- **On-Page** – This is how your content communicates its purpose—clearly, consistently, and in alignment with what the audience is actually searching for. On-page SEO isn't about tricks or tactics. It's about making sure each page is structured in a way that signals relevance, earns trust, and matches real **search intent**. Titles, descriptions, headlines, and structure all play a role in turning content into discoverable, usable, and high-performing assets.

Tool tip: *Search intent.*
The underlying reason behind a query—what the user actually wants to accomplish. Common types include informational (learn something), navigational (find a brand or page), commercial (compare options), and transactional (make a purchase). Matching content to the correct intent is critical for visibility and conversion.

- **Off-Page** – If technical SEO is the foundation and on-page SEO is the framing, off-page SEO is your external reputation—what others say about your brand when you're not in the room. It's built through **backlinks**, brand mentions, citations, digital PR, and other signals of credibility. These references validate your authority in the eyes of search engines and users alike. The stronger your reputation outside your website, the more weight your content carries within it. You can't fake this—it has to be earned.

Tool tip: *Backlinks.*
Links from other websites pointing to yours. High-quality backlinks—from reputable, relevant sources—act as endorsements of your credibility and help search engines evaluate your authority. It's not just about quantity; relevance, trust, and context matter most.

This model is essential for practitioners—the people doing the work. It maps well to how SEOs structure audits, plan campaigns, and report performance. You'll find it broken down in detail in **Book III of the SEO Trilogy**, which focuses on tactical execution and day-to-day SEO workflows.

But for leadership, these three pillars are too limited.

They don't reflect how SEO functions across departments, how it influences conversion, or how it integrates with your company's broader digital strategy.

That's why this book focuses on five interdependent pillars that make SEO work as a system:
Technical, Content, UX, Authority, and Data.

The Five Pillars of an SEO System

Technical

This is the foundation your entire site stands on—crawlability, indexation, internal linking, and performance. Technical SEO refers to the non-content elements that help search engines access, render, and interpret your site efficiently. This includes your site architecture, sitemaps, robots.txt, and performance metrics like Core Web Vitals. If search engines can't easily fetch and understand your pages, everything else is handicapped.

Content

Content is how you meet search intent with clarity and depth. That includes not just body copy, but the overall structure of your site—where topics live, how they connect, and how clearly each page addresses a specific purpose. Search intent is the underlying reason behind a query—whether the user is looking to learn, compare, navigate, or transact—and it should shape the angle, depth, and format of your content. Titles, headings, multimedia, and structured data all play a role in satisfying that intent.

UX (User Experience)

Design and flow either help users complete the job they came to do—or get in the way. User experience encompasses layout, readability, mobile behavior, accessibility, and how easily users can take action (e.g., forms, CTAs, checkout). These UX factors directly influence engagement signals like time on page, bounce rate, and task completion, which in turn correlate with better visibility and higher conversion rates.

Authority

Search engines and people both rely on trust signals to evaluate credibility. Authority is built through real-world validation—press coverage, expert authorship, consistent NAP (name, address, phone) data, and high-quality references. It's not just about backlinks. Brand mentions, partnerships, reviews, and bios that reflect genuine expertise help search engines assess whether your content is worth ranking—and help users decide whether to take action.

Data

You need reliable instrumentation to steer the system. That means analytics that separate branded from non-branded demand, dashboards that map keyword intent to actual landing pages, and governance that ensures your reports stay honest and useful. Data governance includes aligning metrics to business outcomes (like leads, revenue, and pipeline), defining sources of truth, and maintaining consistent tracking practices across campaigns, platforms, and teams.

These pillars are interdependent. Strong content on a slow, disorganized site won't scale. Fast pages with thin, misaligned content won't convert. A beautiful UX without authority won't surface. And without measurement, you cannot defend investment or iterate with confidence.

Understanding the five pillars is essential—but **understanding alone doesn't build momentum**. Strategy becomes meaningful only when it's implemented through real workflows, by real people, across departments that often aren't trained to think in SEO terms. That's why building a strong foundation takes more than theory. It takes time, clarity, and coordination. The following example illustrates what that looks like in practice.

7. Building a strong SEO foundation

When I begin working with a new company, I don't come in swinging. I start with caution—**learning, observing, and understanding** how the organization works. I study how teams operate, how decisions are made, and where SEO lives—or more often, where it's been forgotten.

While doing that, I quietly begin fixing technical bugs. Indexation issues. Crawl traps. Redundant schema. Bloated page structures. These are the invisible friction points that hold a site back, and cleaning them up often takes the better part of a year.

But that first year isn't just about code. It's about **building bridges**.

I identify and connect with the people who need to be part of the SEO system—whether they know it yet or not. Content writers, copywriters, web developers, video editors, analytics teams, brand managers, PR, legal, UX, photography, paid media, social media—**almost every department has a role to play in organic growth**. They just haven't been shown how.

That's one of the reasons this book exists: **to help leadership see that SEO is already part of everyone's job—even if no one's told them.**

Once that foundation of trust is established and the technical ground is stable, I begin pointing to the real opportunities—**not what *could* move the needle, but what *will*.**

By then, I've earned the room—and more importantly, I know exactly where the system is leaking value.

Through experience, I've:

- **Restructured entire URL architectures** to improve user navigation and give search engines a clearer understanding of our content strategy.

- **Consolidated bloated or redundant categories**, merging thousands of legacy URLs to build topical authority and eliminate cannibalization.

- **Authored reputation management playbooks** to elevate off-page authority and improve how users interact with the brand across the web.

- **Implemented custom schema markup**—not plug-and-play scripts, but structured data tailored to our actual content and goals, avoiding the bloated noise of software-generated schema.

- **Controlled enterprise-level crawl budgets** with strategic XML sitemap directives that prevent search engines from wasting resources on non-essential pages.

And that's just a handful of examples.

Over the years, I've executed **hundreds** of high-impact SEO initiatives. Some are technical. Others are strategic. But all of them reinforce the same idea:

When SEO is embedded into your system—not just your marketing plan—visibility is no longer a guessing game. It becomes a competitive edge.

SEO doesn't fail because the theory is wrong. It fails because teams don't understand how their piece of the puzzle affects visibility, engagement, or revenue.

Why Companies Fail—Even with Great People in Place

It's a common misconception:
If we hire the strongest SEO agency, we should be confident in our ROI.
If we bring in a highly paid SEO professional, we should be able to turn things around.

The problem isn't the strategy.
The problem is **implementation and quality control—** and who is responsible for it.

Even the best agencies will meet you exactly where your **statement of work (SOW)** ends.
Your internal team must do the rest. And if the people responsible for publishing content, updating the website, managing templates, or handling media **don't have a solid grasp of SEO principles**, the strongest strategy will still fail.

This is why organizations often feel frustrated after bringing in outside experts or consultants:

- The recommendations make sense.

- The plan is clear.
- But the results don't come.

It's not because the strategy was flawed—it's because **execution was fragmented.**

When SEO becomes a shared responsibility—supported by leaders, understood across teams, and backed by process—it moves quickly and scales well. But when it's treated as a service someone else "owns," it becomes a stalled initiative buried under internal friction.

The truth is: **SEO doesn't live in the deck. It lives in the details.** And unless those details are handled correctly by the people inside the company, no outside expert can make it work on their own.

Here are a few real-world examples that illustrate how easily things can go wrong—**and why executive oversight matters.**

1. A blog filled with content—but no direction

A major media site had invested heavily in publishing. The volume of content was impressive—but traffic and engagement lagged behind competitors. On closer inspection, the problem was clear:

- Articles were **not aligned with any search intent**—no clarity on who they were for or what queries they answered

- There were **no clear calls to action**—nothing prompting the reader to explore, convert, or subscribe

- Internal links often pointed to unrelated topics, diluting both SEO value and user experience

- **Heading structures (H1, H2, etc.) were missing or misused**, making the content harder to parse— for readers and search engines

- Articles were duplicated across multiple categories, leading to **keyword cannibalization and diluted authority**

- Critical category pages were mistakenly tagged "noindex," **removing them from search visibility entirely**

To leadership, the blog looked productive.
To search engines, it looked incoherent.

Without someone connecting content production to SEO outcomes, all that effort was underperforming.

2. A well-intentioned video strategy that hurt performance

Video is often touted as a driver of engagement—and it can be. But one client saw traffic **drop** after implementing on-page videos.

Here's what went wrong:

- Videos were added directly to the site **without a delivery platform** (like YouTube or a proper content delivery network). As a result, **load times increased significantly**, hurting performance metrics.

- The videos were placed **below the fold**, meaning search engines didn't prioritize them, and users often missed them entirely.

- There was **no structured data markup** to indicate that video content was present, so it didn't enhance search listings or drive visibility.

This wasn't a bad idea—it was a lack of quality control in the implementation.
SEO isn't about adding features. It's about integrating them **the right way**.

3. High-end imagery that nearly tanked discoverability

Another organization wanted to create a visually stunning experience. They prioritized high-resolution photography to deliver a "premium feel." But they didn't optimize the images.

Each page loaded with multiple **multi-megabyte files**, slowing the experience and hurting mobile performance.

After properly compressing and optimizing the images (without sacrificing quality), and applying structured image metadata, the result was immediate:

- Within two weeks, the site captured **80% of the market share** for relevant image searches on Google in their category.
- No new content. No new campaigns. Just **precision in execution**.

Each of these scenarios had something in common:

- Talent was present.

- Investment had been made.

- The problem wasn't visible—until someone with SEO expertise surfaced it.

That's why **executive understanding matters**.

SEO problems are often silent. They don't trigger error messages. They don't show up in dashboards—until they cost visibility, rankings, and revenue.

When SEO is left to chance, even smart teams can make decisions that hurt performance.

But when it's **understood at the leadership level**, you start to build a system where content, performance, UX, and visibility work together—not against each other.

The SEO ownership map: how growth happens across teams

Search performance doesn't live in one department—it's the outcome of dozens of decisions made across product, content, design, development, marketing, and analytics. That's why SEO often underperforms: not because no one owns it, but because everyone touches it, and no one aligns around it.

This map reframes SEO as a cross-functional system. Each category represents a critical layer that contributes to organic visibility, from how pages are structured to how they're discovered, shared, and understood. For each one, there is a clear **accountable owner**, key collaborators, and a leadership role: to ensure alignment, resource the right work, and remove the friction that stalls progress.

You won't find checklists or tactics here. Instead, each category highlights:

- **Why it matters** to the business

- **Who owns it** and what decisions leaders must support

- **Common issues** that quietly cost visibility or growth

- A clear, narrative-driven **strategy** to align with your SEO team

- The expected **business outcome** when that area is working well

This structure is designed to help leaders understand the real mechanics of SEO—not just rankings, but the infrastructure, clarity, and coordination required to drive sustainable growth. When each of these areas is supported by the right team, equipped with the right tools, and aligned to a shared outcome, SEO becomes not just a marketing channel—but a strategic advantage.

Crawlability & indexation

Why it matters: Search engines can't rank what they can't reach. When high-value pages become hard to find or unintentionally hidden, traffic drops—and no amount of content or design fixes it. This category protects discoverability at the system level.

Accountable role: Web Development Lead
Key collaborators: SEO Lead, Analytics, Content

Decisions & resourcing:

- Approve a clear indexing policy (what should and shouldn't appear in search).

- Fund a basic release safeguard to prevent SEO-critical errors.

- Support a response framework for fixing indexing problems quickly.

Common issues:

- Key pages accidentally blocked from search or buried in navigation.

- Duplicate URLs competing with each other.

- Redirect chains or broken links after updates.

Strategy to align with your SEO team:
Treat your most valuable pages—those tied to lead generation or revenue—as protected assets. Agree on a short list of these priority sections and build them into your release process: they must always be included, reachable within two clicks, and free of duplication or redirect clutter. Schedule a monthly check-in to confirm that nothing has slipped. This isn't about chasing

technical perfection—it's about making sure your growth-driving pages are always visible and working as intended.

Business outcome:
When crawlability is managed deliberately, you avoid invisible losses. Your most important pages remain consistently discoverable, your SEO investments compound over time, and when something does break, the team catches it early—before it costs you traffic or leads.

PDF assets

Why it matters: PDFs often hold valuable content—guides, brochures, forms, specs—but they're rarely treated as strategic assets. If they're invisible to search engines, hard to use on mobile, or disconnected from the rest of the site, they can drag down performance and miss conversion opportunities.

Accountable role: Content Lead
Key collaborators: SEO Lead, Web Development, Design

Decisions & resourcing:

- Approve a governance model for public-facing PDFs (naming, placement, and formatting).

- Ensure teams responsible for creating or uploading PDFs follow SEO standards.

- Fund periodic cleanups to retire or replace outdated files.

Common issues:

- PDFs buried behind clicks, unlinked, or blocked from being indexed.

- Documents missing titles, metadata, or branded templates.

- Outdated versions still discoverable in search.

- No clear connection from the PDF back to the site.

Strategy to align with your SEO team:
Treat every public PDF like a landing page. Apply naming conventions that match real searches, add meta titles and document properties, and include branded headers and links back to key pages. If a PDF ranks in search, it should feel like a deliberate entry point—not an orphaned attachment. Pair this with a simple inventory review every quarter to remove duplicates or outdated versions. With a small effort, your existing PDFs can support both discoverability and trust.

Business outcome:
Your PDF library becomes searchable, measurable, and aligned with your goals. Instead of losing visibility to overlooked documents, you gain new traffic streams, stronger brand signals, and a better user experience—all from content you already have.

Domain trust signals

Why it matters: Your domain sends constant signals about legitimacy, stability, and ownership. Search engines rely on these signals to decide whether to trust, show, or suppress your content—especially when multiple sites or brands are involved.

Accountable role: IT or DevOps Lead
Key collaborators: SEO Lead, Legal, Security

Decisions & resourcing:

- Maintain clear ownership and access to all domains and subdomains.

- Approve redirect policies that consolidate authority across properties.

- Support regular audits to prevent trust and security gaps.

Common issues:

- Legacy domains still indexed or showing outdated content.

- Unsecured (HTTP) pages or mismatched redirects.

- Subdomains or microsites competing with the main site.

- Domain ownership info that's inconsistent, expired, or hidden.

Strategy to align with your SEO team:
Treat your domain and subdomains as a single, managed ecosystem. Maintain an active inventory of what's live and why, with a clear policy on what should redirect, stay active, or be retired. Ensure every domain reflects a secure and unified brand—via HTTPS, consistent WHOIS records, and clean redirects. If you've acquired other domains or sunset older sites, work with SEO to protect and pass on any authority. For international or product-specific subdomains, validate that they're supporting—not splitting—your search presence.

Business outcome:
A clean and trusted domain structure builds credibility

with search engines and users. You prevent fragmentation, reduce security risks, and improve the impact of every campaign by keeping trust signals aligned. This gives your brand a stronger foundation for long-term visibility and growth.

Google Search Console

Why it matters: Google Search Console (GSC) is the source of truth for how Google sees your site. It reveals what's being indexed, how your pages appear in search, and where technical or content issues may be holding back performance. If your team isn't using it—or doesn't have access—they're flying blind.

Accountable role: SEO Lead
Key collaborators: Analytics, Web Development, Content

Decisions & resourcing:

- Ensure all verified properties are active and accessible to relevant teams.

- Approve workflows that route key GSC insights (errors, coverage issues, keyword shifts) to owners.

- Support regular monitoring—not just when traffic drops.

Common issues:

- Key sections or subdomains missing from GSC entirely.

- Ownership tied to personal accounts or lost credentials.

- Indexation problems identified in GSC but unresolved due to lack of clarity or follow-up.

- Content decisions made without insight into how pages are actually performing in search.

Strategy to align with your SEO team:
Make GSC part of your normal review rhythm. Verify all active properties—main domains, subdomains, and international versions—and assign ownership to a business account, not individuals. Define a core set of reports to check monthly: index coverage, search performance, page experience, enhancements, and manual actions. Empower your SEO team to translate those findings into action, and ensure each stakeholder knows which GSC insights affect their part of the site. Don't treat GSC as a troubleshooting tool—treat it as a real-time visibility dashboard.

Business outcome:
With the right access and review process in place, GSC becomes an early warning system and a source of opportunity. Teams can catch issues before they impact traffic and make smarter decisions based on what's really working in search. It keeps your SEO efforts focused, measurable, and aligned with how Google sees your site.

Internal linking

Why it matters: Internal links shape how search engines understand your site—and how users navigate it. They signal which pages matter most, guide the flow of authority, and connect related content. A smart internal linking structure turns scattered content into a cohesive system that ranks better and converts more.

Accountable role: SEO Lead

Key collaborators: Content, Web Development, UX

Decisions & resourcing:

- Support linking standards for content teams (e.g., every new page links up and down the hierarchy).

- Approve design patterns that surface key links (in navs, modules, and footers).

- Fund structural fixes when deep or isolated pages aren't getting discovered.

Common issues:

- High-value pages buried too deep or with no internal links pointing to them.

- Editorial teams publishing without a linking strategy.

- Broken or outdated links after site changes.

- Overuse of generic anchor text that misses context.

Strategy to align with your SEO team:

Think of your site as a map. The homepage and core sections should naturally guide visitors—and search engines—to important, relevant destinations. Ask your SEO team to identify which pages are underlinked and create a plan to surface them more often in navigation, hubs, or content modules. Set expectations that every new page supports the structure—not just lives in isolation. Even simple link modules (e.g., "Related topics" or "You might also like") can strengthen visibility and help users explore deeper.

Business outcome:
With a clear internal linking strategy, important pages get found faster, rank higher, and keep users engaged longer. You reduce reliance on external links to perform well in search and improve both discoverability and conversion—just by making smarter use of the content you already have.

Internationalization

Why it matters: If your business serves multiple countries, languages, or regions, search engines need clear signals about which pages to show where. Without proper international setup, your content can compete against itself, confuse users, or simply fail to appear in the right markets.

Accountable role: Web Development Lead
Key collaborators: SEO Lead, Localization, Legal

Decisions & resourcing:

- Approve the structure for multilingual and multi-regional content (e.g., subfolders vs. subdomains).

- Support the implementation of language and region signals across all versions.

- Ensure ongoing updates follow a consistent international model—not just local patches.

Common issues:

- One language version showing up in the wrong country's search results.

- Inconsistent use of localized metadata, design, or legal disclaimers.

- Missing or broken hreflang tags (hreflang tells search engines which version of a page to show in each language or region).

- Separate teams creating conflicting or duplicative content without coordination.

Strategy to align with your SEO team:
Create one global framework—and stick to it. Decide on the ideal site structure for international growth and document how each version of a page should be created, tagged, and linked. Work with your SEO team to ensure hreflang is implemented correctly and systematically across all pages. If localization is outsourced or handled by regional teams, enforce consistency with a shared SEO and UX standard. International SEO isn't just about translation—it's about precision in signaling.

Business outcome:
With a strong internationalization strategy, your brand appears correctly in every market you serve. You avoid missed traffic, prevent cannibalization between regions, and build a unified global presence. This ensures your investment in localization actually delivers the reach and discoverability it deserves.

Local SEO

Why it matters: When customers search with local intent—by city, neighborhood, or "near me"—they're usually ready to take action. Local SEO ensures your business shows up when and where it matters most. It's a direct path to calls, visits, bookings, and revenue.

Accountable role: Local Marketing Manager
Key collaborators: SEO Lead, Social, Customer Service

Decisions & resourcing:

- Approve governance for managing Google Business Profiles and location pages.

- Assign clear ownership of review response workflows.

- Ensure accurate and consistent business info across platforms.

Common issues:

- Inconsistent or outdated listings for hours, services, or locations.

- Reviews ignored or handled without brand alignment.

- Location pages with thin or duplicated content.

- No structured plan for growing visibility in competitive local markets.

Strategy to align with your SEO team:
Treat each location like a storefront that must earn its own visibility. Make sure every Google Business Profile is claimed, complete, and actively managed—this includes categories, descriptions, photos, service lists, and links to your site. Your SEO team can help ensure each location has a high-quality, indexable landing page that's optimized for local search terms and connected to internal navigation. Reviews play a critical role here: develop a proactive system for gathering and responding to feedback. Local trust is a ranking factor—and a conversion driver.

Business outcome:

A well-managed local presence drives more foot traffic, calls, and leads from customers in your service area. You increase visibility in high-intent searches, strengthen brand credibility, and support measurable results at the local level—all without relying on paid ads. When done right, Local SEO turns proximity into performance.

Analytics & insights

Why it matters: Without accurate data, there's no way to measure what's working—or what's holding you back. Analytics connects effort to outcomes. It allows your team to prioritize, diagnose, and prove impact over time. For SEO, that means tying visibility to real business performance.

Accountable role: Analytics Lead
Key collaborators: SEO Lead, Content, Web Development

Decisions & resourcing:

- Approve consistent tracking across all domains, subdomains, and environments.

- Ensure teams have access to SEO-relevant dashboards and metrics.

- Support QA processes that catch tracking issues early—especially after site changes.

Common issues:

- Pages missing key analytics tags or firing incorrect data.

- SEO performance only tracked at the channel level—not by page, topic, or intent.

- Confusion over which metrics matter (e.g., clicks vs. impressions vs. conversions).

- Broken dashboards or delays in surfacing trends.

Strategy to align with your SEO team:
Establish a shared understanding of what success looks like in search—beyond just rankings. Set up dashboards that combine data from Google Search Console, Google Analytics, and other SEO tools to track performance at the content level: what pages are being discovered, clicked, and engaged with. Work with your team to create a recurring review cadence (monthly or quarterly) where trends are not only reported but interpreted. Data should drive action—not just sit in a report.

Business outcome:
With the right analytics in place, SEO becomes a strategic lever you can monitor, adjust, and scale. Leadership sees clear signals of growth, content teams get feedback that sharpens future efforts, and technical teams can spot issues before they become costly. Insights turn SEO from a black box into a repeatable, measurable engine for performance.

Image optimization

Why it matters: Images influence everything from page speed to search rankings to user engagement. But if they're too large, untagged, or poorly named, they can slow down performance and miss critical visibility in image search. Optimized images make your pages faster, more discoverable, and more compelling.

Accountable role: Web Development Lead
Key collaborators: SEO Lead, Design, Content

Decisions & resourcing:

- Approve optimization standards for all new and existing images.

- Fund tools or workflows that automate compression, naming, and tagging.

- Support collaboration between design and SEO to align on format and metadata needs.

Common issues:

- Oversized images slowing down page load times—especially on mobile.

- Missing or vague alt text that reduces accessibility and search clarity.

- File names that don't reflect page or keyword intent.

- Duplicate assets or outdated versions stored and served inconsistently.

Strategy to align with your SEO team:

Build a lightweight image governance system. Start with shared guidelines: every image should be compressed without quality loss, named to reflect its content or topic, and include descriptive alt text.

Tool tip: *ALT text*

Short for "alternative text," it helps screen readers describe images to visually impaired users and allows search engines to understand image content for indexing and relevance. Well-written ALT text supports both accessibility and SEO.

Align design and SEO so that each team understands where images appear, what purpose they serve, and how they support performance goals. For larger sites, automate where possible—compression, format conversion, and metadata tagging can often be streamlined.

Business outcome:
Optimized images improve page speed, strengthen content clarity, and expand visibility through image search. They enhance both user experience and discoverability—especially for product, service, and location pages. Over time, this leads to faster load times, higher engagement, and increased qualified traffic without adding new content.

Video optimization

Why it matters: Video can boost engagement, explain complex topics, and increase time on page—but only if it's visible, fast, and indexable. Poorly optimized video hurts performance, slows down pages, and often goes unseen by search engines. Treated strategically, video becomes a powerful driver of discoverability and trust.

Accountable role: Web Development Lead
Key collaborators: SEO Lead, Video Production, Design

Decisions & resourcing:

- Approve video hosting and delivery strategy (self-hosted vs. platforms like YouTube or Vimeo).

- Support structured data and metadata tagging for video content.

- Fund scalable templates for embedding video above the fold with clear context.

Common issues:

- Videos buried low on the page or loaded after user interaction—so they're missed by search engines.

- No schema markup to help video appear in rich results.

- Large file sizes slowing down load times or hurting Core Web Vitals.

- Videos published on third-party platforms without any link or embed on your site.

Strategy to align with your SEO team:

Make video a core element of your content structure—not just a visual add-on. Collaborate with SEO to ensure each video is **embedded in a relevant, crawlable page**, surrounded by context that reinforces its topic. Apply **VideoObject schema** with key properties like title, description, thumbnail, duration, and upload date to help search engines surface it in rich results. Use fast-loading formats, CDNs, or video platforms optimized for performance. And above all: place the video where it gets seen. Visibility drives indexability.

Tool tip: *Video indexation*
For a page to be indexed as a video asset, the video must appear **above the fold**—*visible without scrolling on desktop and mobile. Embeds hidden behind tabs or buried too low may be ignored by search engines.*

Business outcome:

Optimized video increases your presence in search

results, drives deeper engagement, and reinforces brand authority. When discoverable and fast, video becomes a multiplier: boosting SEO performance, improving user experience, and extending the reach of your content across both search and social platforms.

Brand authority & off-page signals

Why it matters: Search engines look beyond your website to evaluate credibility. Mentions, links, reviews, and how people talk about your brand all influence whether your content gets surfaced—or sidelined. Off-page signals build the authority that makes rankings stick.

Accountable role: PR or Communications Lead
Key collaborators: SEO Lead, Partnerships, Legal, Social

Decisions & resourcing:

- Align PR, partnerships, and content strategies to support long-term brand visibility.

- Approve outreach and publishing efforts that grow credible mentions and backlinks.

- Ensure brand and legal teams are not unintentionally limiting earned visibility (e.g., removing links or blocking embeds).

Common issues:

- Inconsistent branding or domain mentions across media and partner sites.

- High-authority mentions that don't link back—or link to the wrong place.

- Link-building handled by third parties with no quality oversight, risking penalties.

- Missed opportunities to amplify high-performing content beyond your own channels.

Strategy to align with your SEO team:
Think of brand authority as a digital reputation score. Encourage teams to collaborate on content and partnerships that naturally earn attention from other trusted websites—media, organizations, local institutions, or industry leaders. Your SEO team can help evaluate link quality, suggest content for amplification, and identify which mentions actually support search visibility. Use PR wins, thought leadership, and social content to grow both awareness and authority—making sure those efforts connect back to your site.

Business outcome:
A strong off-page presence makes your site more competitive in search, especially in saturated markets. It supports long-term ranking stability, improves trust signals, and amplifies the reach of every campaign. Authority doesn't just live on your site—it's built across the web.

Page profile optimization

Why it matters: Every page on your site is a potential entry point from search. When it's clearly structured, aligned with user intent, and easy to understand—for both people and search engines—it performs better. This is the core of **on-page SEO**: optimizing each page's individual profile for visibility and relevance.

Accountable role: SEO Lead
Key collaborators: Content, Design, Web Development

Decisions & resourcing:

- Approve content standards that align with how users search (headlines, metadata, copy structure).

- Support collaboration between content, design, and SEO during page creation—not just after launch.

- Fund iterative improvements based on search performance—not one-and-done publishing.

Common issues:

- Pages missing clear titles, headings, or keyword context.

- Duplicate or outdated pages competing with better content.

- Thin or bloated content that doesn't match what users expect.

- Teams creating content without visibility into how it performs in search.

Strategy to align with your SEO team:

Treat each page as a product with a profile to manage. That profile includes the title tag, meta description, heading structure, copy, internal links, and supporting media—all mapped to a specific search intent. Your SEO team can help define that intent and guide the structure to support both rankings and clarity. Think of **on-page optimization** not as a checklist, but as a way to make every page purposeful and findable.

Business outcome:

With optimized page profiles, more of your content ranks, gets clicked, and converts. You reduce cannibalization, improve discoverability, and drive growth with content you already own—without always needing more.

Page experience

Why it matters: How a page loads, behaves, and responds affects everything from rankings to conversions. Search engines reward fast, stable, mobile-friendly pages—and users do too. Page experience is where performance, design, and usability meet business outcomes.

Accountable role: Front-End Lead
Key collaborators: SEO Lead, UX, Web Development, Design

Decisions & resourcing:

- Approve performance budgets and load-time targets during page design and development.

- Fund fixes for layout shifts, slow interactions, and mobile usability issues.

- Support ongoing monitoring of Core Web Vitals and UX signals post-launch.

Common issues:

- Slow-loading pages that drive bounce before the content is seen.

- Layout shifts that make forms, buttons, or menus frustrating to use.

- Pages that technically "work" but don't meet Google's page experience thresholds.

- SEO efforts undermined by slow templates, heavy scripts, or unoptimized assets.

Strategy to align with your SEO team:
Page experience isn't just about speed—it's about perception. Collaborate with your SEO and front-end teams to ensure that each page meets both technical standards (like Core Web Vitals) and human expectations. Identify your highest-value templates— home, product, lead gen—and prioritize improvements there. Your SEO team can help monitor real-world performance using field data and align optimization efforts with the metrics that matter most for visibility and engagement.

Business outcome:
Faster, more stable pages improve rankings, lower bounce rates, and increase conversions. Optimizing page experience supports both search visibility and user satisfaction—making every click more likely to lead to action. It's not just about meeting benchmarks—it's about earning attention and trust from the moment the page loads.

Quality assurance

Why it matters: Even small errors—broken links, missing tags, incorrect redirects—can quietly erode your search performance. Quality Assurance (QA) ensures that what gets published is structurally sound, discoverable, and free of issues that hurt visibility, user trust, or technical health.

Accountable role: Web Development Lead
Key collaborators: SEO Lead, QA/Test Engineering, Content

Decisions & resourcing:

- Approve pre-launch SEO checks as part of the standard release process.

- Fund tooling or automation to detect technical and structural issues early.

- Ensure accountability for post-launch fixes when quality issues impact performance.

Common issues:

- Pages go live with broken internal links, redirect errors, or missing metadata.

- Changes to navigation or templates unintentionally remove key SEO elements.

- SEO audits catch problems—but there's no defined workflow to resolve them.

- Quality control depends on individuals, not systems.

Strategy to align with your SEO team:
Build SEO into QA, not after it. Define a short list of must-pass checks for any page or feature release: titles and meta tags, header structure, canonical tags, alt text, internal links, and proper status codes. Your SEO team can provide the checklist and flag critical defects. Treat SEO errors like broken functionality—they can affect traffic, revenue, and brand perception just as much.

Combine manual review with automated tools to scale the process.

Business outcome:
When SEO QA is embedded into your release process, performance becomes more predictable and less reactive. You prevent avoidable losses, reduce the cost of fixes, and protect your investment in content and development. High-quality pages aren't just well-written—they're structurally sound and built to perform.

SEO tool enablement

Why it matters: SEO tools are how teams find issues, spot opportunities, and track performance at scale. Without the right tools in place—and properly configured—teams rely on guesswork or find problems only after traffic drops. Tool enablement turns SEO into a measurable, repeatable system.

Accountable role: SEO Lead
Key collaborators: Analytics, Web Development, IT/Security

Decisions & resourcing:

- Approve access and licensing for tools that support crawling, auditing, and performance monitoring.

- Ensure tools are integrated into workflows, not siloed.

- Support the time and ownership needed to act on what the tools reveal.

Common issues:

- Tools are installed but not configured to match site structure or goals.

- Insights get generated but are never routed to the right teams for action.

- Multiple tools report conflicting data without a source of truth.

- Teams rely on manual checks that miss systemic issues.

Strategy to align with your SEO team:
Treat tools as part of your SEO infrastructure—not a plugin or optional layer. Ask your SEO team to identify which tools are mission-critical and whether current setups reflect the full scope of your site (domains, subdomains, dynamic pages). Prioritize tools that automate technical audits, content checks, and keyword tracking—and build workflows that tie insights to action. A tool that surfaces problems is only valuable if there's a process to fix them.

Business outcome:
With the right SEO tools fully enabled, your team can spot issues before they become losses, measure the impact of changes, and continuously improve site performance. Tool enablement creates consistency, saves time, and empowers data-driven growth across technical and content teams.

Social & sharing

Why it matters: While social signals aren't direct ranking factors, they amplify content reach, build brand awareness, and influence how content is perceived and linked across the web. Sharing also impacts how your pages appear when linked on platforms—affecting engagement, credibility, and click-through.

Accountable role: Social Media Lead
Key collaborators: SEO Lead, Content, PR/Comms, Design

Decisions & resourcing:

- Approve a clear brand strategy for how content is distributed across social platforms.

- Ensure all public-facing content supports proper link previews and shareability.

- Support a coordinated effort between SEO and social teams during content launches.

Common issues:

- Pages missing metadata for social previews (like Open Graph or Twitter Cards).

- Broken or unbranded link previews, reducing trust and engagement.

- High-value content published on-site but never amplified socially.

- No tracking or analysis of how shared content performs in search or referral traffic.

Strategy to align with your SEO team:
Social visibility starts with technical readiness. Ensure key pages and blog content include the right metadata so that when shared, they display strong headlines, images, and descriptions. Collaborate across teams to identify high-performing or evergreen content worth recirculating. Your SEO team can help surface these opportunities and

ensure that new content launches include a promotion layer. Social sharing doesn't just build clicks—it builds recognition and relevance over time.

Tool tip: Evergreen content
Content that stays relevant over long periods of time because it focuses on fundamentals, not short-lived trends. Examples include how-to guides, definitions, pillar pages, and FAQs that continue attracting qualified traffic months or years after publication.

Business outcome:

With optimized sharing and coordinated amplification, your content earns more reach, links, and branded engagement. This strengthens your digital footprint, increases referral traffic, and supports long-term SEO performance—especially for thought leadership, product launches, and locally targeted content. Sharing isn't just marketing—it's momentum.

Structured data

Why it matters: Structured data helps search engines—and increasingly, AI systems—understand what your content means, not just what it says. It powers enhanced search results, improves content classification, and ensures your pages are eligible for rich features like FAQs, product listings, events, and more. In an AI-driven search landscape, structured data is essential.

Accountable role: SEO Lead
Key collaborators: Web Development, Content, Legal

Decisions & resourcing:

- Approve the use of schema markup for key content types (products, services, locations, articles).

- Ensure structured data is aligned with actual on-page content and updated regularly.

- Support testing and validation before publishing.

Common issues:

- Missing schema markup on high-value pages.

- Structured data that is incomplete, outdated, or technically invalid.

- Pages marked up in ways that don't match the visible content—triggering warnings or penalties.

- Teams overlooking new opportunities for AI and search engines that rely on structured context.

Strategy to align with your SEO team:
Make structured data part of your publishing standard. Your SEO team can define which schema types best fit your site and help implement markup that describes content clearly—what it is, who authored it, where it applies, and why it matters. As AI engines increasingly rely on structured information to generate answers, summaries, and previews, schema becomes a key input for how your brand is represented beyond your website.

Business outcome:
Structured data increases the accuracy and visibility of your content across traditional and AI-powered search. It unlocks enhanced listings, improves content targeting, and makes your pages easier to understand, index, and feature—whether by a search engine, chatbot, or digital assistant. It's not just for SEO—it's for the future of content discovery.

URL architecture

Why it matters: Your URL structure shapes how content is organized, discovered, and understood—by both users and search engines. Clean, consistent URL architecture improves crawlability, reduces duplication, and makes SEO more scalable across teams and platforms.

Accountable role: Web Development Lead
Key collaborators: SEO Lead, Product, Content

Decisions & resourcing:

- Approve URL naming conventions and enforce them across CMS and product systems.

- Support redirect policies and URL changes tied to rebrands, launches, or migrations.

- Fund updates that clean up legacy or fragmented structures.

Common issues:

- Poorly formatted slugs (uppercase letters, spaces, underscores, special characters) or auto-generated URLs that reduce clarity and hinder indexing.

- Parameter-heavy or faceted URLs creating infinite crawl paths.

- Poorly planned changes that break links, orphan pages, or lose traffic.

- Key sections buried too deep in the structure to be prioritized in search.

Strategy to align with your SEO team:
Treat URLs as permanent addresses. Work with your SEO team to define structure rules—lowercase, hyphenated, short, and intentional. Ensure naming aligns with how users search and how search engines group

relevance (e.g., /services/consulting vs. /page?id=32). For large sites, plan and test redirect strategies carefully. A well-structured URL system supports clean navigation, faster indexing, and stronger internal linking—all of which impact visibility and growth.

Business outcome:
A clear and consistent URL architecture strengthens your site's foundation. It allows new content to scale without chaos, makes key pages easier to find and rank, and reduces long-term maintenance costs. When URLs are treated strategically—not just technically—they become one of the simplest, most effective levers for sustainable SEO.

XML sitemaps

Why it matters: XML sitemaps act as a roadmap for search engines, guiding them to your most important content. While not a guarantee of indexing, sitemaps improve the odds that high-value pages are discovered quickly and consistently—especially on large or complex sites.

Accountable role: Web Development Lead
Key collaborators: SEO Lead, Analytics

Decisions & resourcing:

- Approve sitemap generation standards that reflect content strategy—not just technical output.

- Ensure updates to key content types (products, locations, blog posts, etc.) are reflected in the sitemap in real time.

- Support submission and monitoring through Google Search Console and other tools.

Common issues:

- Important pages missing from the sitemap, or low-value pages being included by default.

- Stale sitemaps that don't reflect new content, retired URLs, or structural changes.

- Separate sitemaps (for images, videos, etc.) not implemented or not submitted.

- Sitemap errors detected but not resolved—hurting crawl efficiency.

Strategy to align with your SEO team:

Make sitemap management part of your publishing system—not an afterthought. Your SEO team can help define which content types belong in the sitemap and which should be excluded. Work with development to ensure automation reflects these rules, especially for dynamic sites. Submit and validate sitemaps regularly through Search Console, and treat errors or coverage gaps as issues worth prioritizing. A sitemap isn't just a file—it's a signal of what matters most on your site.

Business outcome:

Well-structured, up-to-date XML sitemaps help search engines find your most important pages faster—especially after launches or large updates. This accelerates discovery, supports better index coverage, and improves the return on your content investments by making sure your best pages don't get overlooked.

Takeaways — Part I: AI and the new SEO landscape

- **SEO isn't dead—it diversified.** AI engines now synthesize answers and reward clear structure, authority, and machine-readable markup. If machines can't read you, you're invisible—even if you "rank."

- **GEO, AIO, and AEO are leadership issues.** Treat Generative/AI/Answer Engine Optimization as the new visibility framework: structure for citation and summarization with schema, authorship, and Q&A patterns.

- **Structure beats slogans.** Quiet, system-wide improvements to schema and architecture compound because they clarify meaning for users and machines.

- **Algorithm updates are business updates.** Observe first, don't panic. If fundamentals are strong, hold; act when gaps persist beyond ~30 days or competitors surge. Reinforce E-E-A-T where trust is critical.

- **E-E-A-T is operational, not cosmetic.** Real authors, credentials, and sources raise trust signals that influence whether you surface or get cited. Make author pages baseline.

- **SEO is a system, not a silo.** Results come from five interdependent pillars—Technical, Content, UX, Authority, Data—implemented with quality control across teams.

- **Quality control prevents silent losses.** Common failure modes: buried/blocked pages, heavy unoptimized media, missing schema, thin blogs, weak internal linking, and missing release-time SEO QA.

Next steps for leaders and business owners

- **Set expectations by phase.** Ask your team to report progress in sequence: impressions, clicks, and conversions. Early wins show visibility (impressions), followed by engagement (clicks), and finally measurable outcomes (leads or sales).

- **Align content to business goals.** Approve topics tied to specific products, markets, or seasons. Require a one-line "why it matters" for every major page, and ask which business goal it supports.

- **Fund templates, not one-offs.** Direct your team to fix page templates—like product, article, or location pages—so improvements scale across the site. Ask which templates, once fixed, will lift the most pages.

- **Insist on machine readability.** Require unique titles, one H1, clear headings, internal links, and structured data on every page. Ask whether machines can easily identify what the page is about and why it's trustworthy.

- **Make answers quotable.** Ensure high-priority pages include short, direct answers to common questions. Ask what key statement or section an AI engine would likely quote if summarizing your page.

- **Reduce friction with simple governance.** Approve a short release checklist—title, meta description, canonical tag, valid schema, and internal links. Ask what safeguards prevent pages from going live without these essentials.

- **Coordinate with paid media.** Request a clear view of where paid campaigns overlap with organic visibility and where they fill gaps. Reinvest overlap into high-value, non-branded opportunities that expand reach.

- **Run a cadence that ships.** Keep review meetings short and focused on three points: last month's impact, what ships this month, and current risks or dependencies. Always close with clear next steps, owners, and dates.

PART II: SEO as a business asset

8. Organic growth: scale without ad spend

Most companies lean on paid media because it's fast and measurable—but the moment the budget stops, so does the traffic. SEO works differently. It takes longer to build, but once it's in place, it becomes the most cost-effective growth engine in your marketing mix.

Unlike paid, organic search compounds. Every improvement—technical, structural, or content-related—strengthens the entire system. You're not buying clicks; you're building discoverability.

Still, many leaders treat SEO as a silver bullet—something that can quickly fix visibility. Others assume it's free because there's no per-click cost. But SEO is neither quick nor free. It takes time, cross-functional alignment, and expert execution across content, development, design, and analytics. **Typically, it takes six to twelve months to see steady, meaningful results.**

And when SEO appears to underperform, it's often not the strategy that's broken. It's the measurement, the execution, or interference from other channels.

SEO vs. paid media: long-term vs. short-term

Paid media is fast and controllable. You set a budget, place an ad, and traffic arrives—until the budget stops. It's a powerful tool for launches, promotions, and testing messages. But paid media is a rental. You don't own the visibility. The moment you stop paying, your presence disappears.

SEO is slower at the start because you're building assets: purposeful pages, smart internal links, structured data, optimized media, and a reputation trail across the web. Once those assets are in place, your cost per acquisition starts to drop—and your momentum builds.

An optimized article can rank for dozens of queries. A well-structured hub can support dozens of product pages. A properly marked-up video can surface on your site, in search results, and inside AI engines. That's the compounding effect at work.

Tool tip: *Compounding*
Compounding in SEO means each improvement (site structure, content, internal links, performance, credibility) increases the effectiveness of the others. The cumulative impact is greater than any single change.

Here's the pragmatic split I recommend:

- Use paid media to generate immediate visibility on day one.

- Use SEO to reduce your dependency on paid by day one hundred.

The most sustainable budgets treat paid media as an accelerator and a safety net—not a replacement for organic traction. If paid is your primary discovery engine year after year, you're not investing in growth—you're renting visibility that disappears when the budget runs out.

Tool tip: *Dark shadow of paid media*
Paid campaigns can mask weak organic strategy because traffic appears strong while cost per acquisition quietly rises. When budgets tighten, the shadow lifts—and the gap in organic performance is revealed.

The Hidden Cost of Competing With Yourself: SEO vs Paid Media

I've seen this firsthand. On one of our most recent projects, we improved technical structure, optimized content, and expanded our organic footprint. As a result, visibility grew from **23.2 million to 39.6 million impressions**.

But clicks didn't follow. In fact, they **dropped—from 267,000 to 242,000**.

To uncover what was happening, we ran a **Dark Week test**, temporarily pulling back on paid search. The results were clear: **paid media was intercepting roughly 65% of our branded organic search traffic**.

What does it mean that paid media was intercepting organic traffic?

When someone searches for your company name or a branded product, paid ads typically appear first—labeled as "Sponsored." Your organic result may still rank at the top, but it's **pushed down**, and the click goes to the ad instead. That's a click you likely would have earned without paying for it.

The paid media budget had grown to **$210,000 per month**, yet a large portion of that spend was being used to compete with our own organic listings. We weren't gaining new visibility—we were **paying to intercept ourselves**.

Tool tip: *Intercepting traffic*
Interception happens when paid ads appear above your organic listings—especially on branded searches—capturing clicks you would have received organically. This inflates paid performance and masks SEO's actual contribution.

This isn't a paid vs. organic debate. It's about alignment. When teams operate in silos, budgets get wasted and insights are lost. But when **paid, SEO, content, and analytics come together**, companies can identify invisible overlaps, streamline spend, and grow more efficiently.

This was a prime example of what happens when SEO is treated as a system, not a department. By educating stakeholders and collaborating with teams who didn't realize they had a direct impact on organic performance, we surfaced over **$136,000 in monthly savings**—without reducing traffic or sacrificing results.

That's the real ROI (return of investment) of SEO leadership: building cross-functional awareness that drives smarter decisions across the entire marketing ecosystem.

The invisible return on investment executives overlook

Organic performance rarely announces itself with a single, dramatic spike. It shows up in:

- Lower blended cost per acquisition over quarters, not weeks.

- Higher conversion rates from visitors who arrived via "solution" or "problem" queries.

- Stronger performance of every other channel (email, social, direct) because users first met you via search and now return.

- Rising **non-branded search** demand for your offerings and categories.

Tool tip: *Non-branded search*
These are queries that don't include your company or product name ("best payroll software for contractors"). Growth here is the clearest signal that you're winning discovery, not just defending your brand.

Because these returns accumulate across many touchpoints, they're easy to miss if your reporting only shows last-click conversions or focuses on weekly fluctuations. Leaders see the ad platform dashboard and feel in control. They don't see the hundreds of low-friction, high-intent paths that organic creates: help articles, comparison pages, local landing pages, buyer's guides, glossary terms, and frequently asked questions, all working together.

Two practical signals I ask leadership to monitor:

1. **Assisted conversions from organic**
   Even when organic isn't the last click, it often starts the relationship. If you hide these assists, you'll underfund the very system that opened the door.

2. **Share of voice for non-branded terms**
   Choose a manageable set of category keywords

and track how often your pages appear and earn clicks. It's a directional "market share of attention" metric for search.

Tool tip: *Share of voice*
A directional metric estimating how much visibility your site commands for a curated set of category keywords. It is not a vanity rank list; it's a basket-level view of presence.

Why SEO isn't "free"—but it scales better than anything

You don't pay platforms for the click, but you absolutely invest to earn it. "Free traffic" is a myth that sets expectations to fail. Here's what the investment actually looks like:

- **Strategy and structure.** Defining topics, mapping them to intents, building hub-and-spoke architectures, and ensuring your navigation and internal links make sense to humans and machines.

- **Content that answers real questions.** Not just publishing blogs, but producing decision-stage pages, product education, comparisons, and local relevance content that match how people search.

- **Technical fitness.** Crawlability, indexation, page performance, structured data, and media handling (images and video) so search engines can parse and present your information with confidence.

- **Credibility signals.** Reviews, references, bylines, press mentions, and helpful external resources.

You're demonstrating expertise and trust, not gaming links.

- **Conversion readiness.** Headlines, calls to action, forms, and page flows that guide users to the next step. Traffic without action is a cost center, not an asset.

Once these pieces are in place, the unit economics get better every quarter:

- **Marginal cost per additional click declines.** A strong page can rank for many related queries; each incremental query you capture adds traffic without proportional spend.

- **Content becomes "owned media."** Updates and improvements build on prior work instead of starting from zero.

- **Cross-channel lift appears.** People who discovered you organically return via direct and email, costing you nothing to reacquire.

Contrast that with paid. Paid has a **flooring cost**: every new click costs at least the auction price. If your conversion rate dips or competition spikes, your cost per acquisition rises—immediately. With organic, your floor is the investment you already made; your ceiling is how effectively you continue to improve.

A practical growth model: how SEO and paid media should evolve together

Instead of thinking in terms of "channels," think in phases—because SEO and paid media aren't in competition. They play different roles at different times.

Here's how the most effective organizations manage that balance:

Phase 1 (0–3 months): Drive immediate traffic while building the SEO foundation

- Launch paid media campaigns to generate early visibility and test key messaging.

- Simultaneously fix technical issues, define your keyword strategy, and improve site structure and navigation.

- Start creating or improving critical content assets—product pages, service pages, and high-priority landing pages.

Phase 2 (3–9 months): Publish core SEO assets and monitor performance

- Publish cornerstone content: category hubs, guides, FAQs, and key decision-stage pages.

- Optimize images, internal links, and structured data to ensure search engines can fully understand and feature your content.

- Begin tracking how organic visibility and conversions improve across branded and high-intent searches.

Phase 3 (9–18 months): Expand content and refine the conversion path

- Scale your content strategy by adding supporting blog posts, educational resources, comparison pages, and region-specific pages.

- Begin refining calls to action, page layouts, and forms to increase conversion rates from organic traffic.

- Use paid media more strategically—focus on promotions, competitive terms, or high-converting audiences.

Phase 4 (18+ months): Optimize, expand, and reduce dependency on paid media

- Reinvest in new organic opportunities—new categories, content formats (e.g., video), and geographies.

- Continue improving site speed, schema markup, and user experience to maximize search performance.

- Paid spend becomes more efficient: focused on product launches, remarketing, or short-term gaps—not on propping up discoverability.

This isn't theory; it's how scalable, defensible growth looks in practice. Each quarter adds assets. Each asset reduces future costs. Each improvement strengthens the rest of the system.

What Leadership Should Be Asking About SEO and Paid Media

Executives don't need to master SEO or paid search to lead effectively. But they do need to ask the right questions—and know what a thoughtful answer looks like. The questions below help uncover **budget inefficiencies**, **channel conflicts**, and **growth opportunities** that may be hidden behind siloed dashboards.

This is what leadership should be asking—and what they should expect in return.

1. Are we bidding on keywords where we already rank organically?

What to expect:
A clear breakdown of where paid campaigns are overlapping with organic rankings—especially in positions 1–3. The team should identify if those paid clicks are necessary (e.g., for competitive defense), or if they're **duplicating efforts and driving up cost unnecessarily**.

2. How are we using paid media to fill gaps SEO can't yet cover?

What to expect:
A strategic breakdown of how paid campaigns are complementing—not duplicating—organic efforts. Leadership should see that paid is being used to support early-stage launches, time-sensitive promotions, or competitive keywords where SEO is still ramping up. This ensures that **each channel plays a defined, complementary role** in overall discoverability.

3. Branded search terms are usually easier wins for SEO—how can paid help with highly relevant non-branded keywords?

What to expect:
A clear explanation of how paid campaigns are helping the brand show up for users who are searching for solutions, categories, or services—but don't yet know the company by name. Paid should be expanding reach while SEO gains ground over time.

4. Where could we safely reduce paid spend without sacrificing traffic or leads?

What to expect:
Evidence-based scenarios where SEO is already covering key terms or pages. The team should offer **testing strategies** (like dark week tests or bid suppression on specific terms) to prove that reducing spend in certain areas won't hurt overall performance—and may **improve ROI**.

5. When organic performance improves, do we adjust our paid strategy accordingly?

What to expect:
An actual process. The team should be able to show that when SEO gains traction, **paid campaigns are reviewed and recalibrated**—whether by lowering bids, shifting budget to competitive terms, or reallocating toward awareness campaigns. If no adjustments are being made, **you're likely overspending.**

9. CRO: turning traffic into revenue

Getting to the top of Google is only half the battle. **You can earn thousands of visits a day**, but if those visitors don't take action—don't call, click, sign up, or buy—then your SEO isn't driving business. It's just generating noise.

That's why **Conversion Rate Optimization (CRO)** isn't optional. It's not a separate discipline. It's the other half of SEO—the part that turns visibility into results.

And yet, most companies haven't even heard the term **"SEO CRO Specialist."** It's rarely a job title, almost never a dedicated hire, and often completely missing from strategy discussions. That's a gap—because when no one owns the conversion side of SEO, traffic grows but performance stalls.

Let's break down what CRO really means, how it influences SEO rankings, and why conversion planning belongs in every SEO strategy—not just after the fact.

Why CRO Belongs Inside Your SEO Strategy

Most SEO strategies focus on getting users to the page. Rankings, impressions, and clicks are measured as signs of progress. But in a business context, those metrics only tell part of the story. Visibility alone doesn't create value—**action does**.

That's where Conversion Rate Optimization (CRO) enters—not as a separate function, but as an essential layer of the SEO strategy itself.

Every search starts with a question, a problem, or a need. When your SEO strategy works, it matches that intent and earns the click. But if the user lands on the page and doesn't know what to do next—if the layout is confusing, the messaging unclear, or the call to action buried—**the opportunity is lost.**

This is a common blind spot. Companies invest heavily in content, keywords, and rankings—but neglect the on-page experience. The assumption is that once the traffic arrives, the work is done. In reality, **that's where the real performance begins**.

A page that ranks but doesn't convert isn't successful. It's incomplete.

And worse, that underperformance doesn't just affect your business metrics—it can also hurt your search visibility. Google increasingly evaluates how users engage with your content. If people regularly bounce back to search results after clicking your page, it sends a signal: this content didn't satisfy the intent. Over time, that can cause rankings to drop—even if your technical SEO is solid.

That's why CRO isn't "post-SEO." It's not something to test after the fact. It's a strategic discipline that should be integrated from the start: when you plan the structure of a page, define its purpose, write its copy, and decide what action you want the user to take.

Yet most organizations don't have a defined role or owner for this work. The idea of an **SEO CRO Specialist**—someone who ensures that organic traffic leads to measurable outcomes—is still foreign to many companies. It's not a title that appears in org charts. But it's a role that already exists in practice—someone in marketing, UX, content, or SEO who sees the gap and quietly tries to bridge it.

The companies that perform best in organic search aren't just optimizing for keywords. They're optimizing for what happens after the click. **They treat CRO as an integrated part of the SEO system—not an afterthought.**

That shift in mindset is what turns traffic into results.

What Shapes Conversions—Beyond Keywords

Search rankings bring people in. What happens next determines whether that visibility translates into business results.

Everyone reading this has likely experienced it: you search for a solution—*how to fix a login error, best small business accounting tool, wedding venues near me*, or *how to start intermittent fasting*. You find a promising result, click the link... and land on a page that seems designed to frustrate you.

The content you were promised is either buried beneath popups, banners, and unrelated offers—or it's missing entirely. You're greeted by competing calls to action, vague headlines, and a sales pitch for something unrelated to your original query. The experience is chaotic, disconnected, and hard to trust.

What happens next?
You leave.
You click the next result.
You go where the answer is clear and the path forward is obvious—often within seconds.

This moment is where **most SEO strategies fall apart**. The click was earned through solid keyword targeting and content alignment. But the conversion—the action, the engagement, the outcome—was lost due to poor on-page experience.

And this is exactly where **Conversion Rate Optimization (CRO)** becomes essential. An **SEO CRO Specialist** ensures that traffic doesn't just arrive—it lands in the right place, sees the right message, and takes the right action.

Executives often assume these on-page details— headlines, layout, button placement—are purely creative or subjective. In reality, they are **performance levers** that affect both business outcomes and search engine rankings. Google observes how users interact with your content. If people stay, engage, and convert, those signals reinforce your SEO visibility. If they bounce, the signal is clear: the experience didn't deliver. Google interprets this as a bad user experience—and it could, and most likely will, negatively affect your ranking.

Elements like headline clarity, structure, visual trust cues, and mobile responsiveness aren't about design trends. They're about aligning the **user's intent** with the **business's offer**—without friction.

A strong SEO strategy gets users to the door.
A well-executed CRO strategy invites them in, shows them what they came for, and makes the next step obvious.

Without that alignment, companies don't just lose conversions—they risk losing their visibility altogether.

How Design, Copy, and CTAs Influence SEO Performance

It's easy to assume that design and copy are aesthetic choices—branding decisions meant to reflect taste, tone, or market positioning. But in the context of search engine optimization, they serve a much more practical role: they shape how users behave on the page.

That behavior is what search engines are watching.

Google's algorithms don't just evaluate keywords and technical structure—they evaluate **what users do when they land on your site.** Do they stay and scroll? Do they bounce quickly? Do they interact with your content or click on something meaningful? Those signals—known as behavioral or engagement signals—are used to determine whether your page is satisfying the user's intent.

That means design, copy, and CTAs directly influence your SEO performance in three measurable ways:

1. **User Engagement Affects Ranking**
   Clean design, clear messaging, and a logical page structure reduce friction and encourage users to stay longer. The longer they stay, the more likely it is that your page answered their query—and the more likely it is that Google will continue ranking it.

2. **Conversion Signals Reinforce Content Quality**
   When users engage with your CTA—whether by filling out a form, clicking a comparison, watching a video, or navigating deeper into the site—it signals that the content delivered value. These are indicators of quality that search engines use to differentiate strong pages from superficial ones.

3. **Poor Experience Undermines Trust—Fast**
   If a visitor lands on a page and is met with a wall of ads, disorganized content, or pushy sales copy, they leave. And when that happens repeatedly, your ranking suffers. Search engines don't want to send users to pages that frustrate them.

In short: **the visual and structural decisions made by design and marketing teams influence search visibility whether they know it or not.**

That's why SEO cannot be isolated from user experience—and why copywriters, designers, and developers are key contributors to search performance, even if "SEO" isn't in their job title.

An SEO CRO Specialist works across those roles to ensure that once a page earns the click, it delivers the experience that keeps the visitor engaged—and the performance that earns long-term visibility.

The Trust Curve of SEO Conversion Strategy

Too often, CRO is treated as an optimization phase that begins after the SEO work is done. But in a high-performing system, **conversion isn't a phase—it's part of the page's original purpose.** Pages that rank but don't convert are incomplete by design.

Why Conversion Optimization Often Starts Late—and Why That's a Mistake

Still, in practice, many SEO professionals find themselves working in reverse—**not by choice, but by necessity**. When a new strategist is brought in, they're often stepping into a site with what leadership sees as "final." The design is beautiful. The copy has already been approved. The branding is locked in. Suggesting changes—especially to improve conversion—can feel like touching sacred ground.

That's why many SEO professionals begin with what can be measured. Early wins—improved rankings, increased visibility, sustained traffic growth—build the credibility needed to move the conversation forward. Once the data speaks for itself, the message becomes clear: *"The SEO strategies are working—do you want to take it to the next level?"* At that point, what was once considered untouchable—sacred ground—is now open for discussion, because the value has been proven. Conversion rate optimization no longer feels like a risk. It becomes the next logical step.

Conversion work often happens later than it should, but it's not the end—it's the second round of the same strategy. And when trust is in place, that round moves faster and goes deeper.

When I work with cross-functional teams, I focus less on how the page is ranked and more on how the visitor experiences it. Does it match the searcher's intent? Does the structure make the next step obvious? Is the copy helpful and confident—or robotic and bloated? These aren't superficial details. They're signals—both to the user and to the algorithm—of whether the content delivers.

10.   The right SEO hiring model for your stage

Most SEO efforts don't fail because "SEO doesn't work." They fail because the organization isn't set up to let it work. Whether you hire a **consultant, an in-house SEO, or an agency**, success depends less on their skills—and more on your company's ability to turn their recommendations into action.

Leadership often assumes SEO is something a specialist handles quietly in the corner—when in reality, it only works when the whole company gets behind it. The person you hire may know search engines, but they don't yet know your products, your audience, or your culture. And even if they bring the right strategy, it still falls apart when cross-functional teams—development, design, editorial, legal, video—ignore the recommendations or lack the bandwidth to act on them.

The result? Implementation drags. What could've taken 6 to 12 months now takes 18 to 24. Early momentum stalls, and confidence erodes—not because the strategy was wrong, but because execution was starved.

And sometimes, the problem starts even earlier: the wrong kind of help was hired for the stage you're in. You needed someone to lead and integrate, but hired someone to "do SEO." Or you hired an agency for production when you didn't yet have internal alignment. In both cases, you end up paying twice—once for the initial effort, again to fix it.

This chapter gives you a way out. You'll learn how to:

- Assess your organization's SEO readiness

- Choose the right hiring model for your stage and structure

- Ask better interview and Request For Proposal (RFP) questions that reveal real operators

- Spot red flags before they burn your budget

Whether you're hiring your first SEO or reassessing a struggling program, this chapter will help you make decisions that don't just check a box—but actually create results. And it starts with a hard truth: no one can "own SEO" alone. Not even the expert you hire. So let's talk about making the *right* hire—and setting them up to win.

The Hidden Cost of Hiring Without Readiness

Hiring for SEO feels like progress. You find someone credible, sign the agreement, hand off the tasks—and wait. But months later, nothing meaningful has changed. Rankings didn't move. Conversions didn't climb. Leadership starts questioning the investment.

What went wrong?

In many cases, the problem wasn't who you hired. It's that the company wasn't structurally or culturally ready to let SEO succeed.

I've seen this firsthand.

Before stepping into enterprise SEO, I handled it all myself—spotting issues, running audits, implementing fixes. I knew how the pieces fit together because I had done every part of it. So when I joined a large organization, I expected smoother execution. Bigger team, better process—right?

Not exactly.

I handed off a set of clean, strategic SEO tickets to the development team. Straightforward items: critical directives in the robots.txt file, implementing lazyload, minor markup adjustments. Nothing complex. But when the next sprint rolled around, none of it had been implemented. No follow-up. No questions. Just silence.

These were senior developers—smart, capable, and deeply technical. But no one had taught them how SEO fits into their work. No one raised a hand to say, "I don't get this." They didn't even know what they didn't know. And because they had other priorities to manage, the SEO tasks naturally slipped to the side.

It wasn't just a knowledge gap. It was a cultural one. SEO wasn't seen as core to product delivery—so it didn't get done.

That's when I learned the deeper truth: SEO doesn't fail because of bad hires. It fails when the **organization isn't equipped to support the work**—technically, operationally, and culturally.

Where SEO Hiring Fails

Once the decision to "invest in SEO" is made, most companies focus on who to hire. But what often determines success isn't the person—it's the context they walk into.

Here's where SEO hiring typically fails:

1. They chase deliverables instead of removing blockers.

Leadership asks for more content, backlinks, or audits—expecting results. But if your site structure is broken, your templates are misaligned, or your dev team is over capacity, those deliverables won't perform.

SEO isn't about how much you produce. It's about whether your system is ready to support what you publish.

2. They assume strategy equals execution.

You bring in a seasoned consultant or a specialized agency. They deliver a clean roadmap and prioritized action plan. But six months later, you're still waiting on basic implementation.

Strategy without follow-through becomes shelfware.

3. They scope by time instead of outcomes.

Hiring a contractor for "40 hours per month" sounds efficient—but SEO work isn't linear. Some weeks need sprints, others need approvals or dev support. Time caps can stall high-impact work mid-flight.

Progress should be scoped around what needs to get done—not just how long someone works.

4. They hire a doer when they needed a driver.

You brought in someone to optimize pages—but no one is getting those changes approved, pushed, or prioritized. Implementation is stuck between teams, waiting on decisions or dev resources.

In these cases, companies don't need more SEO tasks. They need someone who can lead change across functions and remove friction.

5. They hire based on ambition, not maturity.

Some companies hire a top-tier agency but have no internal champion to guide implementation. Others bring in a junior hire, expecting them to own technical audits, stakeholder education, and roadmap planning on their own.

The right hire depends on your current reality—not your aspirations. Instead, match your hiring model to your operational maturity:

- **If you're in an early or turnaround stage**, you need speed, clarity, and direction. The best fit is a **senior consultant or a fractional SEO lead—** someone who can diagnose quickly, build a strategy, triage urgent issues, and lay the foundation for sustainable growth.

- **If your company is growing with momentum**, and you already have some content and dev operations in place, it's time to bring in an **in-house SEO** to run the day-to-day. Pair them with a **strategic advisor or consultant** who can guide quarterly priorities, watch for blind spots, and keep the roadmap evolving.

- **If you're managing a complex organization—** multiple brands, teams, or markets—you'll need more capacity. This is where a **specialized SEO agency** makes sense, but only if paired with an **internal SEO owner**. That internal lead ensures the agency's work stays aligned to business goals and gets implemented effectively.

Each of these models can work. But only when the structure around them—support, ownership, resourcing— is ready to let them win.

6. They forget about conversion.

Traffic grows. Rankings improve. But nothing moves the bottom line. Why? Because your pages weren't built to convert.

When SEO is separated from CRO (Conversion Rate Optimization), you end up optimizing for visibility—not results.

7. The cost of hiring too early—or in isolation

Hiring for SEO too early—or without organizational support—creates a false sense of progress. You feel like you're moving forward, but the system underneath hasn't changed.

That's how companies end up paying twice: once for the strategy, and again to fix the gap between recommendations and reality.

Are We Ready to Hire for SEO?

Before bringing in an SEO consultant, agency, or full-time hire, take a step back. Most companies underestimate what it actually takes to execute SEO work—not just at the task level, but across functions. Hiring someone can feel like progress. But if the organization isn't ready to support them, the work stalls, the results delay, and the investment feels wasted.

Use this checklist to assess your readiness:

SEO Readiness Checklist

Check off each item that applies:

> We have executive buy-in to prioritize SEO as a business function, not just a marketing tactic.

> We know what success looks like (beyond traffic—think leads, sales, or conversions).

> We can identify internal owners across development, content, creative, UX, analytics, and legal.

> Our dev team has bandwidth to implement SEO recommendations within a reasonable sprint cycle.

> We can publish or update content without excessive approvals or delays.

> We have a basic analytics setup to track SEO performance and measure change over time.

> We include conversion rate optimization in our SEO roadmap—not as a separate initiative.

> We can commit to a monthly release cadence—even if small—to keep momentum.

> We're open to training cross-functional teams on how SEO relates to their role.

If you can't check at least five of these boxes, your organization may not be ready to bring in execution-focused SEO help just yet.

That doesn't mean you shouldn't start—it means your first hire should be someone who can **build the system**, not just work within one.

What to Do If You're Not Ready

If you're missing key elements from the checklist, the next best step isn't to hire a content writer or sign with a production-focused SEO vendor. It's to bring in an experienced **SEO Strategist** or **Senior Consultant**—someone who can:

- Evaluate your current state

- Identify what's missing or misaligned

- Prioritize what actually moves the needle

- Prepare your teams (or train them) to implement SEO correctly

- Create a roadmap that scales with your business

This role bridges the gap between theory and execution. They help you build the muscle before you try to run.

Please Note: The Audit Will Always Come First

No matter who you hire—whether it's a consultant, an in-house SEO, or a full-service agency—an **audit and competitive analysis will be one of the first things they do**.

This is not a sign that previous work was wasted. It's standard, necessary, and often non-negotiable. Every practitioner needs to verify the current technical health of the site, understand how your pages are performing, and re-establish a baseline they trust. Expect it, and budget for it.

Also, be aware that **not all audits are created equal**. Some are surface-level and fully automated. Others are deep technical evaluations that look at templates, crawl efficiency, schema structure, on-page cannibalization, internal link flow, content depth, and conversion friction.

The deeper the dive, the more actionable the insights— and the higher the cost. But that cost often saves much more downstream.

Not All Agencies Work the Same

If you're considering working with an agency, it's critical to understand how different models operate. **Not all SEO agencies implement what they recommend.** Some only guide. Others act like embedded teams. The gap between expectation and delivery is one of the biggest reasons agency relationships fail.

Here's a breakdown:

1. Strategy & Advisory-Only Agencies

These firms focus on diagnosis and recommendations. They provide detailed audits, roadmaps, keyword research, and content strategies—but stop short of hands-on implementation.

- **What they deliver:** Reports, prioritization frameworks, content and technical guidance.

- **What they expect from you:** Your internal dev, content, and design teams to do the actual work.

- **Best fit for:** Organizations with strong internal bandwidth and SEO-literate teams.

If your team is already stretched thin, advisory-only agencies may give you great plans—but little actual progress.

2. Full-Array SEO Agencies

These are operational partners that cover everything from strategy to execution. They bring their own team—writers, technical SEOs, developers, designers, and analysts—and act as an extension of your marketing or product org.

- **What they deliver:** Everything from audits to content production, on-page optimization, template adjustments, structured data, CRO, and reporting.

- **What they expect from you:** Access, approvals, and alignment—so they can move quickly.

- **Best fit for:** Companies that want impact but don't have the team or time to build it internally.

A full-array agency often functions like a plug-and-play SEO department. If implementation is where you've historically stalled, this is the support model to look for.

3. Hybrid Agencies

Some agencies offer a hybrid approach: they'll own strategy and partial implementation (like content production or technical fixes), but still rely on your internal teams for things like deployments, final QA, or stakeholder approvals.

- **Best fit for:** Companies with moderate internal capacity and the ability to collaborate closely with outside partners.

Know What You're Buying

Many SEO engagements fail not because the agency underdelivered—but because no one aligned expectations upfront. Did you want strategy or execution? Do you have the internal support to implement what's recommended? Are you ready for weekly feedback loops, or just quarterly reviews?

The more honest you are about your readiness, the more successful your SEO partnership will be.

What to Look for in Interviews and RFPs

Hiring the right SEO partner—consultant, in-house, or agency—requires more than evaluating credentials. You're looking for someone who can diagnose clearly, navigate your internal complexity, and tie their work to measurable outcomes.

Whether you're running interviews or issuing an RFP, focus on **capabilities**, not just past job titles.

The Three Capabilities That Predict Success

These are the capabilities that separate SEO tacticians from true operators. Whether you're interviewing a candidate or evaluating an agency, these are the qualities that determine whether the work will actually get done—and drive results.

1. Diagnostic clarity

Can they cut through the noise and pinpoint what matters?

Great SEO operators don't flood you with data—they identify what's broken, what's missing, and what's worth fixing first. They understand that not all issues carry equal weight, and they can explain the *why* behind their prioritization.

Ask them to review one of your live pages. In five minutes, can they surface the top three blockers and recommend two high-leverage experiments—without needing to run a full audit first?

You're not just testing technical knowledge. You're testing their ability to **filter complexity into action**.

2. Change management

Can they get work shipped in your real-world environment?

Most SEO projects stall not at the strategy level—but at implementation. A strong operator knows how to work across development, content, legal, design, and leadership to move things forward. They don't just file tickets—they get things done.

Look for someone who can:

- Influence roadmaps and priorities
- Train non-SEO teams (editors, devs, UX) on the "why" behind their asks
- Introduce friction-reducing tools: page briefs, content templates, Definitions of Done
- Follow up and unblock without escalation every time

You're not just hiring an optimizer. You're hiring a **navigator of internal complexity**.

3. Measurement discipline

Can they build a performance framework that informs real decisions?

SEO produces a lot of data. A strong operator knows which metrics to watch early (leading indicators) and how those tie to business results (lagging indicators). Just as important, they know how to translate those numbers into next steps.

Tool tip: *Leading vs. Lagging Indicators*
Leading: Early signals like index coverage, keyword footprint growth, or template-level Core Web Vitals.
Lagging: Final outcomes such as conversions, Marketing Qualified Leads (MQLs), assisted revenue, or share of organic traffic.

They should be able to explain:

- Which signals indicate early progress (e.g., indexed pages, query expansion, Core Web Vitals improvement)

- Which business metrics matter most (e.g., qualified leads, assisted revenue, page-level conversion)

- How they've adjusted strategy in past roles based on what the data showed

Great operators connect the two—and act before the lagging data catches up.

Interview Prompts That Reveal Capability

Ask these in real interviews—or include them as scenario tasks in your vetting process:

- **"Walk me through an SEO initiative that improved revenue without publishing more pages."**
  Look for answers around internal linking, UX changes, SERP intent alignment, or offer clarity. You're listening for system thinking—not "more blogs."

- **"Show me a time an algorithm update impacted your site. What did you change—and what did you avoid changing—in the first two weeks?"**
  Look for thoughtful restraint, clear hypotheses, and data-informed responses—not panic reactions.

- **"Pick one of our product or service pages. In 90 seconds, tell me the most important fix and why."**
  You're testing for prioritization under pressure—not a feature dump.

- **"How do you ensure SEO and paid media don't cannibalize each other?"**
  Strong answers mention branded vs. non-branded segmentation, **query mapping**, and collaborative planning across channels.

Tool tip: *Query mapping or search intent mapping*
The practice of assigning specific search intents (informational, commercial, transactional, navigational) to each page or funnel step—so organic and paid efforts support, rather than compete with, each other.

What to Include in Your RFP (and What to Ignore)

Most SEO RFPs are filled with broad goals, empty checklists, or buzzword bingo. They invite impressive-looking proposals—but not necessarily the right ones. If you want to avoid paying for theater, your RFP needs to reflect the real work of SEO: prioritization, coordination, and measurable progress.

Here's what actually matters—and what separates strategic partners from those who just sound good on paper.

1. Business context and constraints

Don't just ask vendors how they'll "increase organic traffic." That's abstract. Tell them what they're walking into.

- What are your revenue goals, product margins, or lead quality benchmarks?

- Are you in a seasonal industry where timing affects opportunity windows?

- How much access do you really have to development and content teams?

- Is your CMS flexible, or does every update require IT?

The more clearly you define your real-world constraints, the better your responses will reflect **practical strategy**, not ideal-world theory.

2. Diagnosis and prioritization

Instead of asking for a laundry list of ideas, ask vendors to demonstrate how they **prioritize**.

Request a sample 90-day plan built around 3–5 initiatives. Each initiative should include:

- A clearly defined **problem**
- A testable **hypothesis**
- Required **resources** (copy, dev, design)
- **Dependencies** that could delay the work
- A defined **success metric** and how it will be measured

This not only reveals how they think—it shows whether they understand how work actually gets done in a cross-functional environment.

3. Operating cadence

Execution lives or dies by the calendar. If your potential partner doesn't have a cadence, you're buying reaction—not progress.

Ask for a sample monthly or quarterly working rhythm:

- What gets reviewed and when?
- Who's involved in those meetings?
- What decisions are made in those reviews?

This will expose whether they're capable of shipping work on a timeline—or just delivering insights you can't act on.

4. Measurement framework

The right partner should be able to propose a **decision-ready dashboard**, not just a pile of metrics.

Ask them to show a sample dashboard with 8–12 metrics—no more. It should include a balance of:

- **Leading indicators** (e.g., indexed pages, query growth, Core Web Vitals)

- **Lagging indicators** (e.g., qualified leads, MQLs, conversion rates)

- And for each metric: what decision it's meant to drive

This will reveal whether they can **think in outcomes**, not just traffic.

5. Implementation plan

Here's where most SEO projects break down. Everyone agrees on what needs to happen—no one's sure who's actually doing it.

Ask the vendor to specify:

- Who will write dev tickets

- Who will edit templates or content

- Who handles QA before something goes live

- What's expected from your team—and what happens if it's delayed

If their implementation plan is vague or missing, assume this is where your project will stall.

One Test That Cuts Through the Noise

Give every vendor the same "page improvement" mini-brief. Pick a real (but anonymized) URL, outline the page's intent and challenges, and give them 72 hours to return a before/after improvement plan.

You'll learn more from that one exercise than from 40 slides of sales material. The ones who think like operators will stand out fast.

Red flags from "experts" that burn your budget

- **Guarantees of rankings or timelines.** Real operators guarantee process, quality, and communication—never positions.

- **Backlink packages or domain authority boosts.** True authority is earned by credible content, PR, partnerships, and brand signals—not by buying lists.

- **Dashboards stuffed with impressions and vanity charts.** If they can't connect work to qualified traffic, engagement, and conversion, you're buying theater.

- **Instant audits that don't touch templates.** If their "technical plan" doesn't include template-level changes, navigation, or content governance, you'll fix symptoms, not causes.

- **Over-indexing on tools over thinking.** Tools are lenses, not strategies. Beware slide decks that are screenshots of software with no business translation.

- **Paywalls on your own data or content.** You should own accounts, dashboards, schema libraries, and documentation.

- **"We'll handle it" with no ticketing or changelog.** If it isn't documented, it didn't happen—and you can't learn from it.

- **No CRO plan.** Any plan that doesn't address calls to action, page hierarchy, or offer clarity will overpay for traffic.

- **Content mills.** If "content strategy" means more posts with no query mapping, internal linking, or SERP analysis, expect thin results.

Choosing the right model (quick guide)

There's no single "right" way to staff SEO. What matters is choosing the model that fits your company's maturity, resources, and internal structure. The three most common approaches—consultant, in-house hire, and agency—each solve different problems.

A **consultant** (often senior or fractional) brings diagnosis, direction, and governance. They can quickly identify what's broken, build a roadmap, and unlock dev, design, and content bottlenecks. Consultants are best when you need fast wins, clarity, or a reset after a failing initiative. But they aren't meant to operate in isolation. Without internal executors, even the sharpest roadmap stalls.

An **in-house hire** provides continuity. They build institutional knowledge, maintain momentum, and connect SEO to marketing, development, and analytics teams. This model works best when your organization already has a steady roadmap and ongoing production needs. The risk? Hiring too junior. Without authority or strategic depth, the role becomes reactive—a task taker instead of a change driver.

An **agency** offers scale and specialization. With multi-disciplinary teams spanning content, design, and technical SEO, they can move fast and handle large operations. But their effectiveness depends on you having a strong internal owner—someone who can prioritize, approve, and enforce alignment. Without that oversight, activity replaces progress, and reports replace results.

The best setups often blend these models: a consultant defines the strategy, an in-house lead governs execution, and an agency handles production. When these three roles are synchronized, SEO functions as a system instead of a silo.

The internal owner for that system should be what marketers call a **T-shaped professional**—someone with broad literacy across disciplines and deep expertise in one. In many modern organizations, this is also described as a **full-stack digital strategist**: someone who understands how strategy, content, design, technology, and infrastructure work together to drive growth.

Tool tip: *T-shaped professional*
A person with broad cross-disciplinary knowledge and deep expertise in one core area. Sometimes referred to as a *full-stack digital strategist*, this profile bridges creative, technical, and business

teams to turn SEO strategy into measurable impact.

How to structure the engagement

Once you've chosen the right model, success depends on how you structure the work. Many SEO programs fail not because the strategy was wrong, but because execution was treated like a campaign instead of a continuous product. The goal is to create a rhythm of delivery, learning, and improvement.

Start by **defining outcomes, not tasks.** Before discussing audits, content, or backlinks, decide what success actually looks like. For example: *Increase non-branded qualified sessions to product pages by 40% within six months and lift the conversion rate from those pages by 20%.* Clear outcomes keep priorities aligned and results measurable.

Then, **deliver in phases, not presentations.** Progress should be visible through completed work, not just reports. Each phase should include tangible deliverables—template updates, a refined information architecture, rewritten pages, or structured-data rollouts—each tied to a measurement plan that tracks its impact.

Set a steady operating cadence. Monthly reviews should cover roadmap progress, blockers, and learnings. Quarterly reviews should step back to evaluate outcomes, adjust strategy, and choose the next focus areas. This structure keeps SEO moving forward while allowing leadership to stay informed without slowing momentum.

Document everything. Maintain a shared workspace for decisions, tickets, page briefs, changelogs, and post-release results. Documentation is not red tape—it's how you preserve institutional knowledge and make sure progress compounds instead of restarting with every new initiative.

Finally, **align SEO with conversion optimization.** Treat them as one budget line and one workstream. Each initiative should carry a shared hypothesis: how visibility improvements and on-page experience work together to drive the same business metric. When SEO and CRO operate in unison, traffic and conversion stop competing for attention and start compounding results.

The one-page score rubric for hiring

Grade each candidate/vendor 1–5 on:

1. **Diagnosis & prioritization**

2. **Integration & change leadership**

3. **Measurement & decision-making**

4. **Execution plan quality** (tickets, QA, release notes)

5. **CRO literacy** (offers, hierarchy, CTAs, forms)

Anything under 3 on (2) or (5) is a deal-breaker for most companies.

SEO roles and descriptions

1. SEO Director – Aligns SEO with company strategy, secures executive buy-in, and ensures the discipline contributes directly to business growth.

2. SEO Consultant – Acts as a senior advisor who diagnoses issues, designs the roadmap, and governs implementation across teams—often in a fractional or external capacity to accelerate results.

3. SEO Manager – Oversees execution, cross-team coordination, and accountability to deliver measurable outcomes.

4. SEO Strategist – Builds the roadmap, uncovers opportunities, and prioritizes actions that align with company goals.

5. SEO CRO Specialist – Converts SEO-driven traffic into business growth by optimizing UX, offer design, and on-page experience; ensures conversion goals are integrated into SEO strategy.

6. SEO Project Manager – Manages timelines, resources, and communication to keep SEO initiatives organized, documented, and on schedule.

7. SEO Account Manager – Manages client relationships and reporting, translating SEO strategy and performance into business outcomes.

8. SEO Specialist – Executes on-page optimizations, audits, and tactical SEO improvements within the roadmap.

9. SEO Technical Specialist – Ensures site crawlability, indexation, and structural health through technical audits and code-level solutions.

10. SEO Content Specialist – Bridges keyword strategy and content production, creating briefs and aligning pages to search intent.

11. SEO Copywriter – Writes optimized, engaging copy that meets both ranking and conversion goals.

12. SEO Video Specialist – Enhances video discoverability, metadata, schema, and performance across search platforms.

13. SEO Image Specialist – Optimizes images for speed, accessibility, structured metadata, and search visibility.

14. SEO Web Developer – Implements technical and structural SEO updates, including schema, redirects, and quality assurance.

15. SEO Analyst – Monitors performance, interprets data, and builds actionable dashboards to guide decision-making.

Your hiring need is based on your SEO roadmap

You don't need to hire every role—but as your SEO demand grows and stakeholders reach capacity, bringing in the right specialists becomes essential. Each role should remove friction, create momentum, and expand your ability to execute. When you staff for **system success**, not just deliverables, SEO turns into a compounding asset.

Your roadmap—not a generic org chart—should dictate who you hire next. Each addition must address a specific constraint—technical, creative, or strategic—and move measurable outcomes forward.

In my case, I operated as the **SEO Manager**, responsible for holistic strategy, prioritization, and cross-team execution. The approach worked, but over time I became the bottleneck. To scale results, I focused on roles that directly supported the next phase of the roadmap.

The immediate priorities were **conversion** and **visibility expansion**. A **CRO Specialist** was essential to turn growing traffic into measurable revenue. A **Video Specialist** added a new discovery channel—boosting visibility in Google Search, Discovery, and YouTube while improving engagement signals on key pages.

If additional resources had been available, the next hires would have been an **SEO Content Specialist** and an **SEO Web Developer**. The content specialist would have accelerated query mapping, brief creation, and consistent publishing. The web developer would have reduced friction from dependency on a centralized development queue that required multiple approvals and delayed key fixes.

Hire for **leverage**, not for titles. Every role should exist to remove a blocker identified in your SEO roadmap and advance the next performance metric.

Here's a tighter, publication-ready cut that removes redundancy and keeps the leadership cadence.

Takeaways — Part II: SEO as a Business Asset

Organic growth: compounding without ad spend

- Paid buys attention; SEO builds equity that lowers acquisition costs over time.

- Align paid and organic to stop branded-term cannibalization and budget leakage.

- Use paid as acceleration while SEO establishes durable discoverability.

- Judge ROI by blended efficiency, non-branded demand, and cross-channel lift.

Conversion rate optimization: where visibility becomes value

- Traffic without action is waste; CRO is the second half of SEO.

- Clear structure, focused copy, and purposeful CTAs improve engagement signals that sustain rankings.

- Design for conversion from the start; don't retrofit it later.

- Treat SEO and CRO as one system: visibility creates opportunity, conversion captures it.

Hiring SEO right

- Failure is usually organizational, not tactical—hire for system ownership.

- Match model to maturity: consultant for clarity, in-house for continuity, agency for scale (with a strong internal owner).

- Evaluate operators on diagnosis, prioritization, and change leadership—not tool lists.

- Make CRO non-negotiable in scope and measurement.

Leadership lens

- Ensure paid and organic coordinate, not compete.

- Track non-branded share of voice and assisted conversions to see compounding impact.

- Consider conversion an SEO metric.

- Staff to remove bottlenecks; every role should increase system throughput.

Core insight

SEO becomes an asset when it compounds—technically, operationally, financially. Paid can buy attention; SEO earns it and turns it into lasting equity.

Next steps for leaders and business owners

1) Clarify the signal

- Ship a simple dashboard: branded vs non-branded, assisted conversions, landing-page CVR, paid-vs-organic overlap.

- Audit and suppress branded ads where you rank 1–3; measure CPA impact.

- Review top landing pages: one purpose, one action, one visible proof.

2) Establish ownership

- Appoint an empowered SEO lead to coordinate dev, content, design, and paid.

- Adopt a **Definition of Done**: implemented, measured, documented.

3) Align and test

- Use paid to cover gaps and validate messaging; rebalance quarterly.

- Track compounding indicators: non-branded share of voice, assisted conversions.

- Integrate CRO into every SEO initiative.

4) Staff for maturity

- Early: senior consultant to diagnose and build the system.

- Growth: in-house lead for continuity and iteration.

- Complex: specialized agency + strong internal owner.

- Scope by releases and outcomes, not hours.

5) Institutionalize improvement

- Maintain a changelog of launches and results; review monthly.

- Run a steady operating cadence: monthly ship review, quarterly strategy reset.

- Treat SEO as a product—each release strengthens the system and compounds results.

PART III: The hidden layers most executives miss

In every successful SEO initiative, the technical foundation is the first layer to be addressed. Not because of an obsession with code, but because every decision that follows depends on it. Once technical issues begin to surface, a site's true maturity becomes clear. Broken architecture, missing tags, and poor performance aren't simply "developer problems." They're business problems that quietly limit how much visibility a company can earn.

The challenge is that most web developers know how to build a website—but not how to make it discoverable. They understand functionality, not visibility. A perfectly designed site can still fail to reach its audience. When that happens, leadership often blames marketing instead of recognizing the disconnect between development and SEO.

Across organizations, the most effective teams are those that understand *why* a fix matters—not just *what* to fix. Once developers and stakeholders grasp the impact of technical decisions on organic visibility, performance changes dramatically. The difference between a compliant site and a competitive one often comes down to that shared understanding.

This section isn't about teaching technicalities—that's covered in Book II of this trilogy. Instead, it focuses on *why* these layers matter. Crawlability, UX flow, performance, and content structure aren't just technical checklists; they're indicators of operational health. When these foundations are weak, strategy never compounds. When they're strong, everything else—content, authority, and conversion—grows faster and more predictably.

Executives don't need to understand every line of code. But they do need to recognize that technical SEO keeps every other effort working as intended. It's the framework that allows creativity, strategy, and data to perform as one. Without it, even the best ideas remain invisible.

11.   Technical SEO, simplified

Most websites don't fail because of bad design. They fail because search engines can't properly crawl or understand them.

Technical SEO is not about tweaking code for the sake of optimization—it's about ensuring the site is *accessible*, *interpretable*, and *performant* enough for both users and search engines to engage with it.

When a site can't be crawled or indexed, no amount of great content, branding, or backlinks can make up for it. It's like hosting a beautiful event in a locked building.

Crawlability: Can Search Engines Reach It?

Search engines work through links. They follow pathways from one page to another, gathering data about what each page contains.

If your website's structure hides pages behind complex navigation, blocked scripts, or dynamic URLs that aren't linked internally, those pages might as well not exist.

From a leadership standpoint, crawlability is your visibility pipeline. Every uncrawled page is lost investment—design time, content production, and user experience that never sees the light of day.

Common crawlability barriers include:

- JavaScript-dependent navigation with no HTML fallbacks

- Internal links hidden in images or scripts

- Disconnected content hubs (no internal linking between related pages)

- Robots.txt or meta tags unintentionally blocking critical URLs

Good crawlability isn't about having a large site—it's about ensuring every page that matters can be reached, read, and understood.

Indexation: Can It Be Stored and Ranked?

Crawlability determines whether a search engine *finds* a page; indexation determines whether it *keeps* it.
Search engines only index pages they consider valuable and unique. Duplicated content, thin pages, or poorly optimized metadata can lead them to skip indexing. Executives often misinterpret this problem as a content issue when it's actually structural. A website with 10,000 pages may have only a few hundred indexed because the rest aren't clear, connected, or useful from a search perspective.

In other words, indexation is not a guarantee—it's an earned privilege.

Broken Architecture: When Design Undermines Discovery

Many teams treat design and SEO as separate disciplines. That separation is often where the cracks begin.
Overdesigned websites—those heavy with animations, oversized images, or fragmented layouts—tend to collapse under their own visual weight. The site might look stunning on launch day, but its code may be unreadable to crawlers, its pages slow to load, and its structure inconsistent.

From an SEO maturity standpoint, architecture is more than visual flow. It's about how pages interconnect, how internal links distribute authority, and how easily users (and bots) can navigate without confusion.
A strong architecture acts like a city grid: predictable, connected, and optimized for movement. A broken one is a maze.

Why Even Pretty Websites Fail in Google

Aesthetics can deceive. Leaders often approve redesigns based on how the site *looks* rather than how it *functions* for discovery. But search engines don't care about color palettes, typography, or animations—they care about efficiency, structure, and relevance.

When developers prioritize appearance over accessibility, performance suffers. And when marketing teams focus on messaging without understanding technical constraints, even the most persuasive content remains buried.

A common symptom: after a redesign, rankings drop, impressions fall, and executives wonder why the "new and improved" website performs worse than before. The answer usually lies in lost crawl paths, missing metadata, and overwritten SEO signals.

Site Health ≠ Design Quality

Healthy websites are not always the best-looking ones— they're the ones that communicate clearly with both humans and machines.
A truly optimized site balances technical precision with creative expression. It's light enough to load quickly, structured enough to be understood, and flexible enough to adapt as search evolves.

12.   UX/UI isn't just design: it's SEO

For years, user experience (UX) and user interface (UI) were treated as creative disciplines—important for aesthetics, branding, and engagement, but separate from search optimization. That separation no longer exists.

If you pay attention to what Google rewards, one pattern becomes undeniable: everything revolves around user experience. Rankings are simply the reflection of how well a site serves its visitors.

Search engines now interpret behavior. They measure how users interact with a page—how long they stay, what they click, and when they leave. A good experience is no longer optional; it's part of the ranking system itself.

UX as a Ranking Factor

Every algorithm update in the past decade has moved closer to rewarding usability. When pages load slowly, confuse visitors, or hide key information, those negative signals tell search engines the page failed to satisfy intent.

UX, in this context, is not just about pleasing visitors—it's about proving *usefulness* at scale.
Metrics such as dwell time, interaction delay, and return visits help algorithms determine whether a page deserves to rank higher. The better the experience, the stronger the ranking signals.

From a leadership perspective, UX is the moment where brand reputation meets measurable performance. A frictionless, intuitive interface increases both satisfaction and conversion, reinforcing SEO through behavioral data.

Visual Hierarchy and Mobile Flow

Visual hierarchy defines how information is presented and consumed. The order, spacing, and emphasis on elements like headlines, buttons, and images guide both users and crawlers through the page.

In the era of mobile-first indexing, this hierarchy matters even more. Mobile visitors navigate with limited patience and screen space. If key information, such as the main headline or call-to-action, is buried below fold or hidden behind sliders, engagement drops—and so does visibility.

A well-structured layout helps search engines interpret importance:

- The *H1* communicates topic intent.

- *H2s* and *H3s* organize supporting context.

- Clear CTAs reinforce purpose and reduce abandonment.

In other words, a page's layout doesn't just impact usability—it tells search engines what matters most.

Bounce Signals and Behavioral Feedback

Bounce rate is often misunderstood as a sign of failure. In reality, it's a behavioral clue. When users land on a page and immediately leave, it usually means the experience didn't match their expectations—most often because of slow load performance or misaligned content, not poor design.

Consider two common scenarios. A page may align perfectly with the user's search intent but take five seconds to load, turning relevance into frustration. The visitor leaves—not because the information was wrong, but because the experience felt like time wasted. In another case, a page might load instantly yet offer little value, context, or depth. The user departs just as quickly, signaling that the content failed to satisfy intent.

Search engines can't read human satisfaction directly, but they infer it. If a page consistently drives users back to search results, it suggests the answer wasn't found. Over time, this erodes ranking potential.

Optimizing UX/UI reduces those negative feedback loops. A clear path, relevant visuals, accessible design, and even embedded video keep users engaged long enough to convert—sending stronger "success" signals back to search engines.

How to Guide Users (and Bots) Through a Page

Guiding users and bots through a page relies on the same principle: structure communicates meaning. A page designed with intent uses visual flow and semantic cues to clarify what's important and what action should follow.

For humans, that means intuitive navigation, contrasting CTAs, and scannable content. For bots, it means proper heading hierarchy, descriptive alt text, and clean internal links that connect related ideas.

When both audiences—human and algorithmic—can move effortlessly from entry to action, SEO ceases to be a technical concern and becomes a user experience strategy.

UX/UI Bottom Line

UX and UI are no longer downstream of SEO—they are SEO. A well-designed interface doesn't just look good; it performs better, ranks higher, and converts faster.

Ignoring usability is no longer a design flaw—it's a visibility loss.

13. Performance is profit

Speed isn't a luxury—it's a revenue condition. In search, slow pages lose visibility; in sales, slow pages lose intent. The outcome is the same: higher acquisition costs and lower conversion efficiency. Performance is the multiplier that makes every other investment—brand, content, paid media—work harder.

What performance really measures

You don't need the mechanics. What matters is whether real users experience pages that load quickly, stay stable, and respond immediately. When that happens, engagement rises and rankings follow. When it doesn't, bounce signals accumulate, discoverability slips, and paid spend starts subsidizing what organic should carry.

The compounding cost of slow

Slowness is a silent tax—paid not once, but every day the site underperforms. It doesn't appear on a balance sheet, yet it compounds across marketing, sales, and customer experience. Every delay costs attention; every lost second costs conversion.

When pages take too long to load, visitors view fewer pages per session, leaving fewer chances to persuade or convert. As engagement drops, ranking signals weaken, shrinking organic visibility. With less organic share, marketing budgets shift toward paid campaigns to recover lost ground—raising acquisition costs and eroding margins.

The pattern is predictable: performance debt becomes financial debt. This is why site speed shows up in finance, not just engineering.

Leadership levers that drive performance

Speed doesn't improve because a developer works harder—it improves because leadership makes it a priority. When performance is treated as a business standard, not a technical aspiration, it becomes everyone's responsibility.

Executives hold the true levers of change. By defining what "fast" means for the brand, setting clear expectations, and enforcing accountability, they determine how the organization values time—both the customer's and its own.

The teams that consistently deliver high-performing websites share one trait: leadership attention. Projects move faster, decisions tighten, and trade-offs become clear when performance is non-negotiable. When the rule is simple—"don't deploy what slows us down"—the entire culture adapts.

It starts with intention.
When leaders make performance part of business governance—reviewed alongside budgets, conversions, and campaign metrics—it stops being a background issue. It becomes a competitive advantage.

Performance excellence isn't achieved through code reviews alone. It's achieved through clear priorities, disciplined launches, and an organization that understands the hidden cost of delay.

Common enterprise anti-patterns

Recognize these before they become line-items in your budget:

- **Overdesign masked as premium.** Autoplay hero videos, oversized imagery, stacked animation libraries—great demo, poor delivery.

- **Script creep.** Heatmaps, chat, tag managers, A/B tools, analytics layers—individually small, collectively slow.

- **One-off landers.** Campaign pages that bypass shared components and re-introduce bloat you already solved elsewhere.

- **Redesign amnesia.** New look, lost routes: broken internal links, changed templates, heavier assets—rankings slip without a clear culprit.

Two leadership scenarios to anchor decisions

- **Relevant but slow.** The page answers the query, but five seconds to first meaningful view turns intent into frustration. The traffic existed—you just didn't get to serve it.

- **Fast but thin.** The page loads instantly, offers little value, and sends people back to search. Speed opened the door; content failed to keep the room.

Both scenarios send the same signal: intent not satisfied. Fixes differ; the business risk is identical.

What to ask—and insist on

- **"Show me the field reality."** What share of sessions meet our experience target on real devices and networks?

- **"Where are the heaviest templates?"** What component or vendor accounts for the top 20% of drag?

- **"What ships next that drops total weight?"** Optimization is a roadmap, not a report— what's the next release that improves load time?

- **"What did we remove?"** Discipline is subtraction. Each month, which scripts/assets did we retire?

The performance bottom line

Performance is not a developer metric; it's a margin lever. Fast pages earn more organic visibility, convert intent at a lower cost, and reduce paid dependency. Treat speed as a standing operating rule, not a one-time project.

Once performance supports discoverability, the next challenge is keeping attention. Speed may open the door, but content determines whether visitors stay. The following section explains why content strategy isn't publishing volume—it's structuring information that search engines can understand and people consider worth their time.

Content Strategy Isn't Blogging

Most companies confuse *content production* with *content strategy.* Publishing frequently isn't the same as publishing with purpose. Strategy defines why content exists, what it's meant to achieve, and how it fits within the broader ecosystem of search, brand, and conversion.

A strong content strategy doesn't chase keywords—it clarifies value. It positions every page to meet an intent, serve a journey stage, and earn measurable visibility. Content production fills space. Content strategy builds equity.

The difference between content production and strategy

Content production answers the question, *"What are we publishing next?"*
Content strategy answers, *"Why are we publishing this, and how does it serve our visibility goals?"*

The distinction matters. A blog calendar without strategy is just noise at scale. Every piece of content should have:

- A defined audience and intent.

- A measurable role in the funnel.

- Internal connections to strengthen topical authority.

- Structured data that signals meaning to search engines.

Without that alignment, even the most active publishing schedule fails to compound results.

When publishing becomes noise

A well-known brand spent thirteen years posting "trending" articles to its blog—more than a thousand pieces. None aligned with the company's products or buyer journeys. The content closed no competitive gaps, strengthened no category authority, and never connected to real search intent. On paper, it looked prolific. In practice, it drowned the brand.

Traffic to the blog increased year after year, giving the illusion of success. The charts went up, and leadership celebrated growth. But the audience it attracted had little to do with the company's market. Visitors came for passing trends, not for products or solutions. Sessions grew while qualified leads, inquiries, and conversions stood still. The brand became popular—but not profitable.

Over time, the hidden costs compounded. Hundreds of irrelevant posts diluted the site's focus, buried important pages, and confused both search engines and readers. Stakeholders resisted change; the thought of deleting years of hard work felt like waste. Yet the real waste was letting that content continue to weigh the brand down. Leadership missed the disconnect because volume felt like momentum.

This is the difference between content production and content strategy. Production fills a calendar. Strategy builds equity and brings conversions.

On-page optimization grading

To help leadership see what "good" looks like, this book introduces a grading framework that quantifies the quality of on-page SEO without requiring technical depth. It evaluates the visible and structural elements that influence discoverability, clarity, and engagement— everything from metadata to design flow.

The framework considers sixteen factors, each weighted by impact: from foundational elements like **Page Title** and **H1 Tag**, to engagement-oriented aspects like

Internal Linking, **FAQ Sections**, and **Conversion Focus.**

The goal isn't to score perfectly. It's to surface gaps. Executives don't need to know how to fix a schema markup or rewrite an H2—they just need to know *why* those elements matter, and whether their teams are treating them as part of a system.

Introducing the on-page quality model

The **On-Page Quality Model** is a diagnostic framework—a scorecard that translates page quality into measurable signals of maturity. It unites structure, intent, and performance into a single evaluation lens, giving leaders a shared way to align creative, technical, and strategic teams around one definition of quality.

It's not a software tool; it's a method of evaluation. Each element—title, meta description, headers, links, schema, and more—represents a *required factor* for visibility. But the presence of these elements alone doesn't guarantee performance. **The quality of their values matters even more.** Titles that fail to match intent, links that don't support relevance, or schema that doesn't reflect the real content all lower the maturity of a page.

In other words, the scorecard measures not just whether SEO components exist—but whether they're implemented according to best practices. Weighted scoring helps identify which elements contribute most to visibility and which are holding it back.

Think of it as a **content x-ray**: it doesn't prescribe what to publish next; it exposes what's preventing your existing content from reaching its potential.

Why the on-page quality model matters for leadership

The scorecard exists because executives need visibility into quality without depending on jargon-filled reports. It helps shift SEO discussions from *volume* to *value.* Instead of "we published ten articles," the question becomes, "how many of them meet the standards of findability, clarity, and intent alignment?"

This shift transforms SEO from a creative exercise into a measurable system of improvement.

Introducing the **On-Page Quality Model**

Element	Revised weight	Visibility Impact
Page title	13 pts	Primary SEO trigger; first thing crawled and shown in SERPs. AEO/GEO: concise, intent-led titles are frequently used as summary/citation anchors.
H1 tag	10 pts	Core on-page indicator of topic alignment; reinforces page focus.
Meta description	11 pts	CTR driver; supports title and

Element	Revised weight	Visibility Impact
		builds trust. AEO/GEO: high-quality summaries often surface as snippet text or seed passages.
Page slug (URL)	7 pts	Affects crawlability, keyword signal, and clean structure.
Main keyword placement (early & natural)	8 pts	Clarifies relevance and ranking context.
Supportive keywords & semantics	4 pts	Strengthens topical depth and entity coverage (useful for broader comprehension, but secondary).
Content depth (query satisfaction)	7 pts	Determines completeness vs. shallow coverage. AEO: improves likelihood of direct answers and rich excerpts.

Element	Revised weight	Visibility Impact
H2–H4 tags (structure & flow)	5 pts	Improves scanability and featured snippet potential. AEO: headings help engines segment and extract answer-ready sections.
Internal linking	4 pts	Distributes equity; clarifies relationships between topics.
External links	2 pts	Adds credibility and context; supports trust signals.
FAQ section	4 pts	Captures long-tail and common questions. AEO: explicit Q&A blocks map cleanly to answer engines.
Schema markup (structured data)	10 pts	Machine-readable context that enables enhanced results. AIO/AEO/GEO: strongest technical signal for meaning,

Element	Revised weight	Visibility Impact
		entities, and relationships.
Asset optimization	3 pts	Enhances UX, accessibility, and crawl efficiency. Includes optimization of images, videos, and downloadable media. Supports **AIO/GEO** by improving how search and generative engines interpret visual and file-based assets.
CTA/conversion focus	3 pts	Drives outcomes; improves engagement signals.
Mobile optimization	3 pts	Foundational for usability and discovery.
Core Web Vitals	3 pts	Real-user speed, stability, responsiveness; supports

Element	Revised weight	Visibility Impact
		engagement and sustained visibility.
Overall UX (design, clarity, flow)	3 pts	Improves dwell time and reduces pogo-sticking.

Tool tip: *Answer engines*
A subset of AI engines designed specifically to deliver synthesized answers to search queries—often directly in the results, without requiring a click. Systems like SGE and Perplexity behave this way. To improve inclusion, use clear structure, helpful content, and schema markup.

The **On-Page Quality Model** turns what was once subjective into something measurable. It exposes where quality exists, where it falters, and where improvement will have the greatest impact. When creative, technical, and leadership teams evaluate content through the same lens, SEO becomes a discipline of clarity—not debate.

Every page on a website tells a story about how well the organization understands its audience, its products, and its purpose. The goal isn't just to publish—it's to communicate relevance so clearly that both humans and machines can recognize value.

With strategy aligned and quality measurable, the next step is to ensure delivery matches intent. Even the most valuable content fails if it isn't fast, stable, and accessible. That's where performance—and every supporting asset—becomes the next layer of advantage.

14. Speed, assets & image optimization

When a company invests heavily in content but neglects its assets, the message rarely reaches its full potential. Every page element—images, PDFs, videos, and even embedded widgets—carries its own SEO weight. Together, these elements can either reinforce your visibility or quietly dilute it.

In many organizations, SEO ends at "publish." Content teams optimize headlines and keywords, but the underlying assets remain heavy, mislabeled, or invisible to search engines. The result: slow discovery, weak engagement, and missed opportunities that no one traces back to the real cause—the assets themselves.

How assets close the invisible gap

Search engines don't just read words; they evaluate how information is delivered. Optimized assets bridge the gap between *what you say* and *how it performs*.

- **A lightweight image** ensures your content loads instantly and keeps users engaged long enough to read it.

- **A correctly labeled file** (with meaningful filenames and alt text) gives search engines precise context—turning visuals into structured, searchable data.

- **An optimized PDF** allows your content to surface as a separate asset in Google results, expanding discoverability beyond the page it lives on.

- **An embedded video with metadata** increases dwell time and adds another indexed touchpoint that reinforces authority.

When those signals align, content stops existing as isolated pages and starts functioning as a connected, data-rich system. That's where compounding visibility happens.

The business value of asset-level SEO

Optimized assets protect the ROI of content creation. Without them, the cost of producing blogs, visuals, and downloadable materials multiplies while reach diminishes.

Asset-level SEO ensures that:

- **Content reaches more channels.** Image search, YouTube, and AI engines increasingly pull from structured media, not just text.

- **Crawl efficiency improves.** Search engines process pages faster when assets are properly referenced, compressed, and cached.

- **Brand signals strengthen.** Consistent filenames, metadata, and captions reinforce your entity identity across platforms.

- **Engagement metrics rise.** Faster-rendering media keeps users interacting longer, improving the behavioral signals search engines track.

Each optimized asset reduces friction, preserves bandwidth, and compounds discoverability. It's the difference between a page that *exists* and one that *performs*.

Metadata: where branding meets performance

Metadata transforms assets from decorative to strategic. With the right details—filename, alt text, caption, and the data embedded inside the file—an image stops being just a visual and becomes a branded, searchable signal that machines can interpret with precision.

Think of metadata as digital packaging. It's the label, description, and provenance your brand carries wherever that image appears. **IPTC fields** such as author, copyright, headline, description, and keywords travel with the asset across platforms, ensuring consistent attribution and reinforcing brand identity. **EXIF data**—like date, device, or location—adds authenticity and can help connect your content to a specific place or event.

There are hundreds of possible metadata fields, but only about twenty truly influence discoverability and SEO value. These are the ones that tell search and AI engines *who created the image, what it represents, where it belongs, and why it matters.* The rest are technical noise.

What's often missed is that **most SEO professionals never touch EXIF or IPTC data.** They stop at filenames and alt text, leaving powerful visibility signals unused. For organizations that invest heavily in visuals—product photography, corporate media, infographics—this overlooked step represents untapped value.

I once orchestrated the optimization of more than **6,000 images in a single day** through batch processing, embedding EXIF and IPTC metadata directly into each file. The result was measurable: those assets captured roughly **80% market share in Google Image Search** for the target category. It wasn't about manipulating algorithms—it was about making every image identifiable, properly attributed, and semantically aligned with the brand's message.

This isn't about becoming a metadata expert—it's about recognizing that every asset your company publishes already has a digital fingerprint. When those fingerprints are optimized, they carry your brand forward across the web. Every embedded field, description, and well-named file strengthens the link between your brand promise and how it's recognized, surfaced, and trusted by search and AI systems.

CDN and DAM: the infrastructure behind asset-level SEO

Behind every fast, consistent, and discoverable brand experience are two unsung systems: the **Content Delivery Network (CDN)** and the **Digital Asset Management (DAM)** platform. They rarely appear in marketing discussions, yet they quietly determine whether your assets perform—or simply exist.

A **CDN** is, in essence, your global delivery layer. It stores copies of your images, videos, and files on servers distributed around the world, ensuring that when someone visits your site, those assets are loaded from the nearest location—not from a single, overloaded origin. This small architectural choice removes the friction of distance. Pages render faster, users stay longer, and search engines reward the experience. For leadership, a CDN isn't a technical luxury—it's infrastructure that protects every marketing investment by guaranteeing that what your teams create actually reaches the customer quickly and consistently.

The **DAM**, on the other hand, governs the creative side of that same ecosystem. It's the central hub where all brand assets live—tagged, versioned, and organized. A good DAM doesn't just store files; it enforces discipline. It keeps filenames consistent, ensures that **EXIF and IPTC metadata** are embedded correctly, and automates how images are resized and distributed across campaigns. Without it, assets get duplicated, misnamed, or stripped of metadata—silently eroding discoverability and brand credibility.

Together, CDN and DAM form the operational backbone of **asset-level SEO**. The CDN accelerates delivery; the DAM preserves meaning and consistency. One ensures your content shows up fast, the other ensures it shows up right. When both are aligned, every image, video, and file your company produces carries the same speed, structure, and clarity as your brand itself—turning invisible infrastructure into visible advantage.

Closing the loop between content and performance

When leadership invests in SEO, content often gets the spotlight—but assets carry the weight. Image optimization, video delivery, and file structure are where the technical and creative sides of SEO meet. These optimizations turn a single page into multiple discoverable surfaces.

For executives, the takeaway is simple: content gets you in the game, but optimized assets keep you visible across every platform that matters. Ignoring them is like funding a billboard campaign where half the posters never get printed.

Speed and metadata are not separate disciplines—they're two halves of the same system. Performance ensures your content is delivered; metadata ensures it's understood. When both align, every asset—whether it's an image, video, or downloadable file—becomes a contributor to visibility, trust, and conversion.

Executives don't need to know how to compress a file or embed metadata, but they do need to make sure it happens. The competitive edge lies in recognizing that these details are not technical extras; they are the connective tissue between creativity and discoverability.

When a company's CDN delivers assets instantly, the DAM enforces metadata discipline, and every image carries the information search and AI engines need to interpret it, the result is seamless brand presence. Pages load fast, visuals surface across platforms, and content investments compound instead of decay.

Most businesses spend heavily on producing content but never optimize how that content travels. The winners will be those that understand speed and metadata not as behind-the-scenes tasks—but as business levers. In SEO, visibility begins long before a search query. It begins with how well your assets are built to be found.

Takeaways — Part III: the hidden layers most executives miss

Technical foundation: where visibility begins
Search visibility depends on infrastructure. If search engines can't access or interpret a site, every marketing and content investment loses reach. Technical strength enables growth across all channels.

UX and UI: the real ranking factors

User experience now drives visibility. Clear navigation, fast load times, and intuitive layouts directly influence rankings and conversions. Good design performs—it doesn't just look better.

Performance as a profit driver

Speed affects revenue. Slow sites reduce engagement, increase ad dependency, and weaken margins. Fast performance compounds visibility and lowers acquisition costs.

Content strategy: clarity over volume

Publishing without intent creates noise. Content earns value when it matches search intent and business goals. Quality, not quantity, determines lasting impact.

Assets and delivery

Images, videos, and files shape how a brand is found and perceived. Optimized, consistent assets extend visibility across platforms and strengthen credibility.

Core insight

SEO maturity comes from system strength, not technical skill. Clean architecture, fast performance, purposeful content, and well-structured assets make a brand easy to find, quick to load, and worth engaging with.

Next steps for leaders and business owners

- **Recognize SEO as infrastructure.** Include it in operational reviews alongside finance, technology, and customer experience—not as a marketing sidebar.

- **Audit visibility health.** Request executive briefings that summarize site accessibility, speed, and content clarity in business terms, not technical jargon.

- **Make performance measurable.** Set company-wide expectations for site responsiveness and user experience just as you would for service quality or uptime.

- **Champion purposeful content.** Approve initiatives only when they serve a defined audience and strategic intent, not just a calendar or trend.

- **Integrate asset governance.** Ask how digital assets are stored, named, and distributed—because each one represents a touchpoint with your brand.

- **Institutionalize review cadence.** Hold quarterly visibility reviews where SEO, marketing, and product leads report on discoverability, engagement, and conversion efficiency.

PART IV: Control, governance & visibility

Most companies think SEO control means owning the analytics dashboard. It doesn't. Real control happens in the unseen layers—the rules, files, and signals that tell search engines and AI systems how to interpret your brand.

I've watched high-performing sites lose visibility overnight, not because their content was bad or their teams were lazy, but because no one was watching the quiet mechanisms underneath. A simple change to a robots.txt file, a misconfigured redirect, or a forgotten sitemap can erase years of momentum. Sometimes it's not even a mistake—just neglect. When no one owns the rules, entropy takes over.

Governance in SEO is like maintaining air traffic control for your digital presence. The pilots—your marketing and content teams—may be skilled, but without someone ensuring that flight paths are clear and signals are accurate, collisions and disappearances are inevitable. Every folder structure, parameter rule, and meta directive forms part of that flight map. Each one determines whether your pages are indexed, understood, and trusted—or lost in the fog of the web.

This part is about regaining that control.
We'll explore how infrastructure and indexing shape what search engines actually see, how schema markup translates your expertise into machine-readable trust, how authority is earned through genuine signals instead of spammy link trades, and how your reputation—both

online and offline—anchors your brand's credibility in search results.

Governance isn't bureaucracy; it's the quiet discipline that keeps growth sustainable. It's the difference between a website that occasionally ranks and a brand that consistently leads.

Control, governance, and visibility are not separate efforts—they are the framework that protects everything you've built so far. Because when the rules are clear, your systems aligned, and your reputation intact, SEO stops being a guessing game and becomes what it was always meant to be: a measurable, controllable business advantage.

15. Infrastructure, indexing & AI governance

Visibility begins where few leaders look: the rules, maps, and manifests that tell machines what a company is, what to crawl, and what to ignore. Content, design, and campaign work can be excellent—and still underperform—when the underlying control layer is missing or mismanaged.

A single directive can move markets. One national network lost a large share of organic traffic after a single line in robots.txt blocked a key directory. No algorithm update, no content problem—just an invisible file changing how the entire site was seen. That is the nature of governance: it is quiet when done well, and painfully loud when neglected.

Infrastructure is the map

URL design, folder hierarchy, and parameter rules establish relationships across the site. When these drift—through redesigns, migrations, or ad-hoc launches—signals fragment. Duplicate patterns emerge, crawl budget is wasted, and equity is split across near-identical URLs. Clear ownership and change control prevent entropy from becoming strategy.

Indexing is permission

Search engines reward what they can fetch, parse, and interpret. That makes robots.txt, XML sitemaps, canonical tags, and redirect policies the gatekeepers of discoverability. Google Search Console is essential, but dashboards are only as useful as the governance beneath them: who owns coverage, who validates exclusions, and who signs off on the "rules of the site."

AI governance is the new perimeter

Beyond traditional search engines, AI systems now crawl, summarize, and learn from web content. Two emerging concepts are shaping how this access might eventually be governed: AI.txt and LLMs.txt.

- **AI.txt** is being discussed as a "robots.txt for AI," a potential standard that would let publishers define what AI systems can access or reuse.

- **LLMs.txt** is envisioned as a companion file—a structured "menu" that directs large language models to short, approved summaries or excerpts instead of full content.

Both are **still proposals, not standards.** To be precise: **as of the third quarter of 2025, neither AI.txt nor**

LLMs.txt has been formally adopted or enforced. They can be explored experimentally, but not relied upon for protection or control.

However, a wave of lawsuits from major media and publishing organizations against AI companies has made one outcome increasingly clear: **the question is no longer *if* a standardized directive will emerge—but *when*.** Legal and commercial pressures are accelerating the push for official mechanisms to govern AI access to online content.

Until that happens, **robots.txt and server-side access controls** remain the only dependable ways to allow or deny crawlers, including AI-related ones.

Strategic posture

Governance is not bureaucracy; it is risk management for revenue. It ensures that architecture stays intentional after launches, that indexation reflects business priorities, and that AI access aligns with the monetization model. Some organizations may choose to block most AI crawlers to protect ad-driven pages; others may publish controlled summaries to encourage attribution and referral. The correct choice depends on business model—not ideology.

Schema & Rich Results

Modern websites are built for people, not for crawlers— and that's often where visibility breaks. Many features designed to enhance user experience inadvertently block what search engines and AI systems can access. For example, **JavaScript functions can prevent content from loading in a way that machines can read**, even if it looks perfect to users. Meanwhile, **schema markup**

does the opposite—it highlights key information in a format search engines understand, increasing the odds that your content is indexed, cited, or featured in results.

Imagine an e-commerce page that invites a visitor to *"Select a model to see price."*
The prices themselves aren't written into the page's HTML; they live in a database. When a user chooses an option, a **JavaScript function** runs behind the scenes, fetching the price and displaying it dynamically.

To a human, that interaction feels instant.
To a crawler, nothing ever happened.

Search engines and AI engines don't "click" buttons, open dropdowns, or trigger scripts. They read the static HTML and structured data that exist before any user interaction. If the price, product name, or description isn't present in that base layer, the crawler can't index or understand it—no matter how well the site performs for visitors. This is why Server Side Rendering (SSR) matters—it ensures key content is delivered in the initial HTML, not hidden behind JavaScript that loads later.

This gap between *what users see* and *what machines see* is one of the most common—and costly—visibility problems. It's why so many sites lose organic traction even with clean design and good UX.

Schema markup fills that gap

By declaring information explicitly in machine-readable form, schema ensures that search engines can "see" the product, the price, the offer, and even the availability—without having to trigger the JavaScript that humans use. It's not a workaround; it's a translation layer.

When structured data describes what the user would see after an interaction, it makes that data **discoverable and indexable**. The product becomes eligible for price snippets, the business for rich results, and the page for more accurate AI summaries.

Think of schema as context at scale:

- It tells machines when a page describes a **person** rather than a company.

- It distinguishes a **service** from a **product**, and a **review** from a random paragraph.

- It defines **FAQs** as helpful, **events** as timely, and **articles** as original sources.

This context is what fuels **rich results**—the enhanced listings that feature images, ratings, FAQs, product details, or snippets at the top of search results. These aren't aesthetic bonuses; they are conversion multipliers. A well-structured page earns both visibility and credibility, signaling to users that the brand behind it is authoritative and transparent.

For companies that rely on organic discovery, schema markup becomes a form of competitive defense. It helps ensure that when AI engines summarize information, your brand remains the *source* of truth—not just another reference. As AI-generated overviews and voice answers increasingly dominate results, structured data is what allows a brand to remain visible in "zero-click" search environments.

However, schema is only effective when used strategically.
Over-marking or using incorrect schema types can

confuse search engines instead of clarifying them. Leadership should view schema as a form of **governance**, not decoration. It represents an intentional commitment to accuracy, transparency, and machine-readable credibility.

In business terms: Schema markup ensures that your company's information is not just published—but understood, trusted, and surfaced at the right moment in the customer journey. At a high level, every visible element on a page has a structured counterpart. When that mapping is intentional, your website stops being a collection of pages and starts functioning as a **knowledge graph** about your company.

On-page element	Structured data relationship	Leadership impact (AIO/AEO/GEO)
FAQs	FAQPage linked to the page's **mainEntity**	Improves topical clarity and captures direct answers for answer engines (AEO); gives AI engines clean Q/A pairs to cite (AIO). Location-qualified FAQs can support local intent (GEO).
Directions & contact info	LocalBusiness, Place, PostalAddress, GeoCoordinates	Ensures assistants and maps return the right address, hours, and phone (GEO). Clear, consistent NAP data also improves machine

On-page element	Structured data relationship	Leadership impact (AIO/AEO/GEO)
		trust and assistant routing (AIO/AEO).
Contact forms & CTAs	Nested under Service, Offer, or Organization (with actionable properties like url, potentialAction)	Makes conversion paths machine-discoverable so assistants can surface "how to contact/book" directly (AEO) and preserve attribution in AI responses (AIO). Localized CTAs reinforce proximity-based actions (GEO).
Authors & experts	Person nested under Article/NewsArticle (with sameAs, credentials)	Strengthens Experience/Expertise signals used by AI and answer engines for source selection and attribution (AIO/AEO). Can pair with local expertise pages when relevant (GEO).
E-E-A-T signals overall	Person, Organization, Review tied to real bios, awards, profiles	Converts trust into durable visibility: assistants prefer credible sources (AIO), answer engines elevate authoritative entities

On-page element	Structured data relationship	Leadership impact (AIO/AEO/GEO)
		(AEO). Local reviews/ratings bolster proximity relevance (GEO).
Content itself	Article, Product, Service, Event, HowTo, VideoObject (declared as **mainEntity**)	Defines the page's purpose so answer engines can select it as the best match (AEO) and AI engines can summarize with correct attribution (AIO). Adding place/region context where appropriate improves local match quality (GEO).

In most cases, **schema markup must be customized**. There are thousands of schema types and properties, and each needs to be chosen and structured to reflect your actual content, organizational model, and personnel. Auto-generated code rarely achieves that precision—it often misrepresents relationships, duplicates data, or omits the context that machines rely on to establish trust.

To ensure full alignment between your visible content and its structured representation, **a Schema expert should review—or ideally write—your markup**. This isn't about adding complexity; it's about protecting clarity. Properly implemented schema turns your website into a consistent, machine-readable reflection of your business—one that search engines, AI assistants, and answer engines can understand, attribute, and reward.

16. Link building vs. authority building

For years, SEO was obsessed with backlinks—the more you had, the higher you ranked. It worked for a while. Then the internet matured, and so did search engines. Today, algorithms evaluate not just *who links to you* but *why* and *how* those signals connect to your brand's overall reputation.

Link building is mechanical. Authority building is relational.
One can be purchased; the other must be earned.

The problem with link chasing

Traditional link-building campaigns often treat links as a commodity: buy a placement, trade a guest post, or post in a directory and call it progress. But search engines have learned to distinguish manufactured popularity from genuine credibility. Artificial links, even when well-disguised, erode trust over time and invite algorithmic scrutiny.

A sudden influx of low-quality backlinks, irrelevant sources, or networked "link wheels" doesn't signal authority—it signals manipulation. Leaders should understand that link volume is no longer a meaningful metric. What matters now is **contextual trust**: whether those links come from relevant, reputable, and authoritative sources that align with your brand's purpose.

Authority in the modern era

Modern authority is built through the same qualities that build human trust: expertise, transparency, and consistency. Search engines measure these qualities through the framework of **E-E-A-T — Experience, Expertise, Authoritativeness, and Trustworthiness.**

- **Experience:** Do you demonstrate firsthand understanding of your industry or product?

- **Expertise:** Is your content written or reviewed by qualified professionals?

- **Authoritativeness:** Do other credible sources recognize or reference your brand?

- **Trustworthiness:** Is your business transparent, accurate, and well-reviewed?

High-authority brands naturally earn links without chasing them. Mentions in the press, citations on credible directories, partnerships, research contributions, and social engagement all compound into a digital footprint that signals real-world trust. Search engines see this footprint as a reflection of reputation, not optimization.

The new signals of authority

Authority today extends beyond traditional backlinks:

- **Reputation signals:** Verified author profiles, consistent company information, and reviews connected through structured data.

- **Citations and mentions:** References on industry publications, events, or local news—links optional, credibility mandatory.

- **Expert content:** Articles, videos, and interviews that earn organic mentions from peers and institutions.

- **Social and PR alignment:** Public relations, brand coverage, and thought leadership that reinforce your topical ownership.

In essence, search and AI engines now use your entire **digital footprint** to measure trust. Backlinks are one small piece of that network.

Leadership perspective

Executives shouldn't ask, *"How many links did we get this quarter?"*
They should ask, *"What are we doing that makes others want to reference us?"*

Authority isn't a marketing tactic—it's the byproduct of real credibility. It comes from subject matter depth, operational transparency, and a consistent message across every channel. High-authority brands don't need as many links because they've already built the trust those links are meant to represent.

If your company has a **Public Relations department**, you already have a goldmine for SEO. PR and SEO share the same foundation: credibility, visibility, and influence. Every press mention, interview, or media feature that PR secures can serve as an *earned authority signal*—one that search engines and AI systems recognize as proof of trust.

When SEO and PR collaborate, authority compounds. PR opens the doors to reputable publications; SEO ensures those mentions are linked, structured, and attributed correctly. Together, they create a measurable lift in brand visibility and trust—both online and off.

For executives, this is one of the most overlooked synergies in modern marketing. Instead of running separate campaigns, align your PR and SEO teams under one shared objective: to make your brand the recognized expert in your space.

17.   Reputation is SEO

Reputation doesn't just influence how people perceive your company—it determines how algorithms evaluate it. Search engines and AI systems are designed to measure trust. They can't read intentions, but they can observe patterns: reviews, mentions, ratings, and how consistently your brand's information appears across the web.

That's why reputation has become one of the most powerful and least understood ranking factors in modern SEO.

Reputation management is not damage control—it's trust architecture

Most people think of **Reputation Management** as crisis response. In reality, it's a discipline that **builds trust before it's tested.** It ensures that every public signal—reviews, press mentions, social discussions, and partnerships—tells a consistent story about your reliability and credibility.

When done right, reputation management becomes a measurable asset. Brands with steady positive sentiment, transparent communication, and verified customer engagement consistently outperform those who ignore it.

These signals don't just attract customers—they attract algorithms.

Search engines interpret this consistency as confidence. They elevate trusted brands higher in results because those brands reduce user risk.

How reputation shows up in SEO signals

- **Press mentions:** Credible media coverage validates your authority in the public domain, amplifying your visibility and trustworthiness.

- **Reviews:** Ratings and customer feedback on Google Business Profile, Yelp, and industry-specific platforms directly influence local rankings and click behavior.

- **Citations and directories:** Consistent business information across directories reinforces your legitimacy. Inconsistencies—like mismatched addresses or outdated phone numbers—erode trust.

- **BBB and industry accreditations:** Third-party trust signals like Better Business Bureau scores or verified certifications are quality indicators recognized by both users and AI systems.

- **Customer service:** Timely, professional responses to reviews—especially negative ones—demonstrate reliability and strengthen perceived trustworthiness.

The Google Business Profile connection

For local and service-based businesses, the **Google Business Profile** (formerly *Google My Business*) is the heartbeat of visibility. It consolidates reviews, contact information, business hours, and real-world proximity into a unified trust signal that Google uses to determine who appears first in local results.

What many executives don't realize is that **ratings influence visibility**.
Profiles with consistently high averages—especially above **4.5 stars**—tend to appear more often and higher in local search and map results. Low or inconsistent review scores can suppress visibility even when other SEO factors are strong. In competitive markets, this difference can decide who gets discovered and who doesn't.

Google's reasoning is simple: when users repeatedly choose highly rated businesses, the algorithm learns that trust and satisfaction predict relevance. A strong reputation becomes a ranking factor.

Accurate information, frequent updates, and genuine engagement amplify that signal. Responding to reviews (positive and negative), updating photos, maintaining accurate hours, and regularly posting updates all reinforce activity and reliability—qualities both users and algorithms reward.

Your profile isn't a listing; it's a live reflection of credibility. The more consistently you nurture it, the more Google treats your business as a preferred result.

PR and SEO: the shared language of credibility

Public Relations and SEO are two sides of the same coin. PR shapes narrative; SEO measures its digital impact. When PR wins credible media coverage, SEO ensures those mentions are linked, structured, and attributed correctly. Together, they create a feedback loop where reputation fuels rankings—and rankings reinforce reputation.

Executives who align PR, customer experience, and SEO strategies create brands that not only rank higher but stay there longer.

Takeaways — Part IV: control, governance & visibility

- **Control lives in the rules, not the dashboard.**
  Visibility depends on quiet mechanisms—
  robots.txt, sitemaps, redirects, schema, and
  structured ownership. When these drift, even great
  content loses traction.

- **Governance prevents invisible loss.**
  SEO failures often come from neglect, not bad
  strategy. Clear change control, defined ownership,
  and documented protocols protect against entropy
  after launches or migrations.

- **Infrastructure defines what machines can see.**
  Architecture, parameters, and indexation rules
  determine which pages exist in search at all.
  Governance translates business priorities into
  technical boundaries.

- **AI access must align with your business
  model.**
  AI.txt and LLMs.txt are emerging—not standard—
  but signal a shift toward publisher-controlled AI
  visibility. Leadership must decide whether AI
  crawlers are partners or risks.

- **Authority can't be bought—it's verified.**
  True visibility grows from consistent, credible
  signals across web, PR, and reviews. Link counts
  fade; trust endures.

- **Reputation is the new ranking factor.**
  Ratings, mentions, and verified identities now
  influence algorithms as much as users. A 4.5-star
  average does more than good design—it drives
  discovery.

Next steps for leaders and business owners

1. **Assign rule ownership.**
    Define who controls robots.txt, sitemaps, redirects, schema, and access logs. Treat these as revenue-critical assets.

2. **Audit technical governance quarterly.**
    Review index coverage, redirect maps, canonical patterns, and AI crawler access. Small misalignments compound quickly.

3. **Elevate schema and reputation to leadership KPIs.**
    Track structured data accuracy and public-review health alongside revenue and traffic. Both now signal brand trust.

4. **Unify SEO, PR, and customer experience.**
    Merge reputation work, earned media, and SEO oversight into one trust strategy. Visibility follows credibility.

5. **Build resilience through clarity.**
    Document every rule, automate monitoring, and align change approvals. SEO stability is now operational governance, not marketing maintenance.

PART V: Executive SEO thinking

18. Metrics that matter: signals that mislead

Executives often see SEO through dashboards—colorful charts, upward arrows, and monthly summaries that look reassuring. But few stop to ask the most important question: *Do these numbers represent real progress, or just movement?*

It's easy to feel confident when impressions climb or traffic spikes after a campaign. The problem is that not all growth is meaningful growth. In SEO, what looks like success can mask stagnation—or worse, distraction.

When visibility sets the stage—but doesn't close the loop

Impressions reflect *discoverability*—how often your pages are eligible to be seen. Strong technical SEO (clean architecture, indexation, speed, structured data) expands that visibility, so impressions often rise first. But impressions aren't outcomes. To turn exposure into traffic, content must align with search intent and communicate value clearly—that's where titles, descriptions, and on-page relevance lift Click-Through Rate. And to turn traffic into revenue, Conversion Rate Optimization—offer clarity, UX flow, trust signals, and friction removal—does the heavy lifting. In short: technical SEO earns the *audience opportunity*, content earns the *click*, and CRO earns the *result*. The executive question isn't "How many saw us?" It's "Are more of the *right* people seeing us, *choosing* us, and *acting*?"

That's where branded versus non-branded traffic becomes essential context. Branded traffic—people searching your company or product by name—reflects awareness that already exists. It's valuable, but it's not new market capture. Non-branded traffic, on the other hand, shows how well your business competes for broader demand—those searching for the problem you solve, not just your logo. Both matter, but they tell very different stories about growth.

When clicks reveal strategic alignment

Click-Through Rate (CTR) is one of the clearest indicators of **intent alignment**—how well your message meets what the audience is actually searching for. When technical SEO strengthens crawlability, page structure, or schema markup, **impressions often surge first** because the site becomes more discoverable. But clicks don't always rise at the same pace. That's because visibility alone doesn't earn the click—**content and message relevance do**. Titles, meta descriptions, and on-page context determine whether users see your page as the best answer to their query.

It's also common for **CTR to dip temporarily** when impressions expand faster than content optimization matures. This isn't failure—it's a **sequencing signal**. Technical SEO opens the door; content optimization persuades the visitor to walk through it. Executives should interpret this lag as progress in motion, not as a performance loss.

But there's another layer leadership must keep in mind: **paid media can intercept organic momentum.** If your paid campaigns are bidding aggressively on the same branded or high-intent keywords your organic pages already rank for, they can suppress organic clicks—even as SEO improves. Searchers may see both results, but ads often claim the first interaction. The data might show declining CTR on organic pages, but the cause isn't poor SEO—it's **channel overlap**.

Understanding this dynamic keeps teams aligned. SEO builds enduring visibility; paid media accelerates exposure—but when they compete instead of complement, they distort metrics and budget decisions. For leadership, CTR isn't just a reflection of message performance—it's a **barometer of how well your channels are integrated**.

When engagement becomes the truth test

Behavioral signals—like dwell time, bounce rate, and pogo-sticking—tell a story most reports ignore. Dwell time measures how long users stay before returning to search results. Bounce rate captures how many leave after viewing just one page. Pogo-sticking, the act of clicking a result, bouncing back, and choosing another, suggests dissatisfaction with what they found.

Executives don't need to calculate these metrics—but they do need to grasp what they represent: **user trust.** A page that attracts visitors but can't hold them signals a gap between promise and delivery. And no metric reveals that gap faster than behavior.

That's why **keyword targeting and content architecture** play such a decisive role in engagement. Many organizations unknowingly compete with themselves—targeting the same topic or keyword across multiple pages. This fragmentation dilutes relevance and confuses search engines about which page deserves to rank. By contrast, creating **dedicated, intent-specific pages** builds topical authority and serves a clearly defined audience need. When content is mapped this way, the result is measurable: visitors find exactly what they were searching for, dwell time increases, bounce rate decreases, and pogo-sticking virtually disappears.

Search engines interpret these engagement improvements as **quality signals**, reinforcing the page's relevance for that intent. But more importantly, these behaviors reflect real human validation—people found value and stopped searching elsewhere. For leadership, that's the ultimate proof that SEO, content strategy, and user experience are working together as one system.

When reports shape perception

Most organizations measure SEO success through what's easy to track, not what's strategically useful. The danger is mistaking activity for achievement—reporting keyword growth instead of conversions, or ranking improvements without audience expansion.

In growth phases, **not every SEO gain shows up where you expect it.** A rise in organic visibility often spills over into other channels—video, PR, social media, or even direct traffic. People discover your brand through search, remember it, and later return without clicking a search result. Those visits look "direct," but they started as organic influence. When leadership understands this connection, they stop chasing isolated metrics and start valuing the compounding visibility SEO creates across the funnel.

Still, **data from real reports matters.** Metrics like keyword expansion, position tracking, clicks, impressions, CTR, domain authority, and engagement indicators—such as dwell time and bounce rate—reveal the *momentum* of your SEO engine. Conversion remains the hardest metric to measure accurately because few customers buy on their first visit. Most will come back through another path once trust has been built. That's not an attribution flaw—it's the way audiences behave.

As a leader, your goal isn't to memorize every metric—it's to interpret what they mean within your business model. Metrics are lenses, not truths. The more you understand their context, the faster you can separate the indicators that **build long-term brand equity** from those that only **decorate reports.**

SEO maturity begins when leadership stops asking, *"What did SEO do this month?"* and starts asking, *"Where is SEO creating durable visibility and recognition across all channels?"* Real progress isn't about how much you can measure—it's about whether your measurement reflects reality.

19.   Aligning SEO with the company vision

You've likely experienced it before: a new product page, campaign landing page, or blog post goes live because a team needs something published fast—yet it's disconnected from brand positioning and audience intent. It ranks for nothing, overlaps with existing content, and fragments authority. Then SEO is invited to "bless it" before launch. That's not strategy; that's rework disguised as progress.

True alignment begins when leadership treats SEO as the **operating system for discoverability**, not a final checklist. SEO must be present from the moment an idea enters production—when new business lines are planned, when a redesign is scoped, or when a market expansion is proposed. Involving SEO early prevents wasted investment, preserves domain authority, and compounds results across channels.

What each team owns (no micromanagement—just clarity)

- **Strategy (Executive leadership + SEO lead)**

    - Define the business outcomes SEO supports (qualified demand, lower Customer Acquisition Cost, brand trust).

    - Approve the SEO roadmap and enforce "SEO first, not last" in every digital initiative.

    - Guard against content cannibalization and ensure clear page ownership.

- **Experience (Content + Design/UX)**

- Align every page to a single user intent and clear action.

- Use consistent structure and hierarchy to strengthen engagement signals.

- Maintain author credibility and current sourcing.

- **Platform (Engineering + Analytics + Legal)**

  - Preserve crawlability, indexation, and performance through every release.

  - Maintain accurate data segmentation (branded vs. non-branded, assisted conversions, return visits).

  - Review early to prevent last-minute changes that weaken intent alignment.

- **Reputation & Demand (PR/Comms + Paid Media + Sales/Customer Success)**

  - Direct coverage and campaigns toward canonical pages that reinforce topical authority.

  - Avoid bidding on high-performing organic terms without a clear strategic purpose.

  - Share voice-of-customer insights to guide positioning and content refinement.

When each group understands its role, SEO stops being a departmental task and becomes a shared performance system. But clarity means little without timing. In many companies, SEO still enters projects after decisions are made—when product names, campaigns, or site architectures are already locked in. At that point, SEO can only **mitigate risk**, not shape opportunity.

This isn't a people issue—it's a process issue. A reactive culture treats SEO as a validator; a strategic culture integrates it as a builder.

That's why the next shift is critical.

Stop "blessing" work at the end: move SEO to the start

Across organizations, SEO remains the last to know about:

- New product or service lines

- New site sections or business pages

- Market entries or acquisitions

- Website redesigns or navigation overhauls

- Domain migrations, rebrands, or CMS transitions

- Analytics, tagging, or privacy platform changes

When SEO is late, strategy turns into retrofit: no content-gap research, no overlap prevention, no authority mapping, no internal linking plan. The fix isn't more meetings—it's a **lightweight governance loop** that brings SEO in at the start of every initiative, ensuring speed without sacrificing strategy.

A simple, durable governance loop

A healthy SEO system doesn't slow teams down; it gives their work direction. The goal isn't to create more meetings—it's to install a predictable rhythm that keeps visibility in focus from concept to release. A lightweight governance loop makes that possible.

Every new initiative should start with a **one-page brief**—a simple document owned by whoever requests the project. It captures the essentials: who the audience is, what problem the page will solve, and how success will be measured. Just as importantly, it names who owns that page and identifies which canonical destination it supports. This early clarity prevents duplication and defines intent before any creative work begins.

Next comes a **pre-flight check**, owned by the SEO lead. Here, potential conflicts are caught before they cost money—overlapping topics, missing internal links, unclear templates, or slow page structures. It's not about perfection; it's about ensuring that what ships will be findable, fast, and consistent with the broader architecture.

The **green-light meeting** is brief by design. In fifteen minutes, teams confirm ownership, delivery dates, and where the new content will be linked from. This is the moment that turns ideas into coordinated action rather than parallel efforts.

After launch, a **post-launch review** closes the loop. Within a few weeks, early indicators—impressions, clicks, and engagement signals—show whether the new page is gaining traction or needs refinement. Paid and organic data are reviewed together to catch overlap and channel conflict before they distort reports.

Why sequencing matters

The timing of SEO involvement determines whether the work compounds or fragments. Before design, pages need a defined purpose so layout serves that goal. Before development, templates must lock in the right structure so engines can interpret them correctly. Before public relations or paid campaigns, canonical targets must be chosen so authority consolidates rather than splits. And before any site migration, redirects and parity checks preserve the equity already earned.

What good alignment looks like

When governance is working, new initiatives launch with a clear owner for each search intent—no duplicate pages competing for the same keyword. Internal links route authority to the right destinations, not the loudest voices. Leadership recognizes that dips in click-through rate during impression surges are signs of progress, not failure. Paid media fills gaps strategically instead of intercepting organic momentum. And post-launch reviews focus on learning and iteration rather than assigning blame.

Leading by Infrastructure, Not Intervention

In the end, SEO governance is not bureaucracy—it's infrastructure for discoverability. Leaders who mandate an **"SEO first, not last"** checkpoint for anything users or search engines will see eliminate most preventable losses. The goal isn't to count pages shipped, but to measure alignment and compounding visibility. Because the cost of late SEO is invisible in the sprint—but painfully visible in the profit and loss statement.

20.   Preparing for the future of SEO

The future of search won't be defined by rankings—it will be defined by relevance and trust across multiple discovery surfaces. The traditional search results page is dissolving into a distributed ecosystem of **AI assistants, answer engines, and contextual feeds** that deliver information long before a user reaches your website. For companies, this shift changes the question from *"How do we rank?"* to *"How do we remain visible and credible in an environment where search happens everywhere?"*

From algorithms to understanding

Search engines are no longer just crawling pages—they're interpreting meaning. Modern AI systems analyze relationships between entities, intent, and authority, not just keywords. Visibility now depends on how well a company structures and expresses its knowledge. Schema markup, consistent terminology, and strong authorship signals help search and AI engines understand your expertise and connect it to evolving queries.

Executives don't need to manage these elements directly, but they do need to **fund the infrastructure** that enables them—structured data, content governance, and integrated analytics that unify insights across channels. In this new era, SEO maturity becomes synonymous with **information architecture maturity**.

When zero-click doesn't mean zero value

Many leaders worry when impressions rise but clicks decline. Yet a growing share of search exposure now happens in **zero-click environments**—featured snippets, AI-generated summaries, voice responses, and predictive suggestions. These moments don't always drive sessions, but they do build **brand familiarity and trust**. When your brand becomes part of the summarized answer, you're not losing traffic—you're gaining authority in context.

Executives should see these surfaces as an extension of their brand presence, not a threat to it. The companies that will thrive are those that **design content to be reusable**—structured so it can appear across search results, voice responses, and AI-assisted experiences.

Infrastructure that adapts

Preparing for this future isn't about chasing every new platform; it's about building **adaptive infrastructure**. Companies need a foundation that allows them to respond quickly when algorithms, devices, or interfaces evolve. That includes maintaining clean site architecture, modular content, consistent metadata, and clear authorship signals.

More importantly, it requires cultural alignment. Teams must think of visibility as a shared outcome—content, design, and development each responsible for ensuring that every page communicates intent clearly to both humans and machines. The organizations that succeed won't be the ones producing more content—they'll be the ones producing **interpretable content**.

Leadership's role in the next chapter

The future of SEO is less about tactics and more about stewardship. Leaders will need to ask new kinds of questions:

- Are we structuring knowledge so that machines can understand our expertise?

- Is our brand discoverable across AI assistants and answer engines, not just search results?

- Are we measuring influence, not just clicks?

Leaders who answer those questions early will shape how their brands are understood—not just how they are found. Because the next era of SEO isn't about optimizing for algorithms; it's about preparing your organization to **be interpreted accurately by intelligent systems**.

When SEO is treated as a **strategic asset, not a tactic**, it becomes a language your business speaks fluently across every platform that touches your audience. That's the kind of visibility that endures—long after algorithms change.

Takeaways — Part V: executive SEO thinking

Metrics that reveal, not mislead

Most dashboards celebrate movement, not progress. Impressions show reach, clicks show relevance, and conversions prove value—but only when read together. Branded vs. non-branded traffic exposes whether growth comes from awareness or true demand. SEO maturity begins when leaders measure cause and context, not decoration.

Integration beats isolation

Technical teams earn visibility, content teams earn clicks, and CRO turns them into revenue. When departments act in sequence, performance compounds; when they work in silos, data distorts. Alignment—not volume—is the multiplier.

Governance is growth control

Bringing SEO in at the start of every digital initiative prevents wasted launches and duplicate pages. A lightweight review loop—brief, pre-flight, and post-launch—keeps visibility intentional instead of accidental.

Infrastructure over intervention

Treat SEO like system architecture, not campaign cleanup. Clean templates, structured data, and cross-channel reporting protect equity and make every release findable, fast, and future-ready.

Relevance is the next ranking

As AI engines and zero-click surfaces expand, visibility shifts from search results to summarized answers. Brands that structure their knowledge and authorship clearly will stay discoverable—even when clicks decline.

Next steps for leaders and business owners

1. **Redefine success**
   Direct your analytics team to build a unified dashboard showing non-branded visibility, assisted conversions, and engagement signals—not vanity metrics.

2. **Adopt an "SEO first, not last" policy**
   Require your product, development, and marketing teams to involve SEO from project start to prevent fragmentation and protect domain authority.

3. **Fund structural readiness**
   Authorize your web and analytics teams to strengthen site architecture, schema, and data integration before scaling new content or campaigns.

4. **Align paid and organic strategies**
   Instruct your marketing and paid media teams to audit keyword overlap and coordinate efforts so paid campaigns support organic growth.

5. **Lead by rhythm, not reaction**
   Schedule quarterly visibility reviews with marketing, product, design, and analytics leads to track progress and remove blockers.

CONCLUSION

• What you now see that most don't

SEO is not a silver bullet. It's not a magic wand you wave today to see results tomorrow. Search engines, AI models, and content platforms don't reward quick fixes— they reward consistency, clarity, and compounding relevance. It takes time for these systems to fully crawl, understand, test, and trust your site over competing options. That's why **momentum is everything**—and delays are costly.

One of the biggest things executives get wrong about SEO is assuming it's a departmental task. It's not. It's an organizational commitment. The SEO team can build the roadmap, monitor performance, and flag risks—but they can't approve content, greenlight changes, or drive cross-functional action. That responsibility belongs to leadership.

If you hold the sign-off, you hold the success. Too often, strategies sit idle for weeks or months, waiting on approval. In the meantime, competitors move forward, and the algorithm recalibrates without you. SEO doesn't fail because it's too complex—it fails because the business never fully commits.

What Executives Get Wrong About SEO isn't just the title of this book—it's the barrier you now have the power to remove. By championing cadence, continuity, and shared ownership, you create the conditions for sustainable growth. You don't need to master the tactics—you just need to clear the runway so the system can do what it was designed to do: perform.

SEO is a system, not a channel.

You've seen how technical health, content quality, user experience, authority, and data all interact. When any one of those elements is weak, the entire system underperforms. SEO isn't one lever—it's a web of moving parts that compound together.

Intent > volume.

Traffic only matters if it matches high-intent queries— those where the user wants to learn, compare, or buy. Chasing keywords without understanding the user's purpose leads to empty sessions and no ROI.

CRO is half the outcome.

If your pages aren't converting, it's not just an SEO issue—it's a conversion issue. Conversion Rate Optimization aligns copy, layout, and calls to action to move users toward real outcomes.

Architecture drives revenue.

A clear URL structure, hub-and-spoke internal linking, and consistent naming conventions make it easier for search engines to crawl, for users to navigate, and for both to convert.

Speed is a financial metric.

Every extra second of load time costs clicks, leads, and advertising efficiency. Core Web Vitals—such as Largest Contentful Paint and Interaction to Next Paint—aren't just developer metrics; they reflect the real user experience and impact revenue.

Authority is earned, not bought.

Backlinks matter, but lasting authority comes from deeper trust signals: press coverage, reviews, citations, bios, and brand consistency. These reflect Experience,

Expertise, Authoritativeness, and Trust (E-E-A-T)—the credibility signals search engines value most.

AI engines are now part of "search."

Answer surfaces, summaries, and zero-click results are shifting how people discover content—often before the traditional results page appears. These systems favor content that's well-structured, informative, and easy to extract into answers.

Structured data is your translator.

Schema markup gives search systems a machine-readable summary of what your page is about. It improves visibility in both classic search results and emerging AI-powered platforms.

Paid media casts a shadow on SEO.

A large ad budget can hide poor UX and weak content—for a while. But it can't substitute for lasting relevance, organic trust, or a discoverable structure. Organic quality is what remains when spend stops.

Dashboards should mirror the business, not vanity metrics.

You now track meaningful signals: non-branded traffic, assisted conversions, content clusters, and page-level outcomes. Average position and impressions are directional—but not decision-worthy in isolation.

Governance beats heroics.

The strongest SEO programs are steady, not reactive. Robots.txt, sitemaps, AI/LLM directives, and consistent deployment discipline protect against avoidable losses. SEO success is often won in maintenance, not emergency sprints.

"Pretty" is not the same as "performant."
Design matters, but only when it supports clarity and action. Aesthetic without function is decoration. The goal is to help users complete the job they came to do—seamlessly.

Scorecards create shared language.
A simple on-page checklist—title, H1, slug, internal links, schema, CTA—helps teams ship pages that are "right the first time," even without SEO experts reviewing every step.

The right hire is a multiplier.
You now look for strategic thinking, cross-functional influence, and the ability to align SEO with CRO, content, development, PR, and analytics. It's not about who knows the most tools—it's who can drive the system.

Organic growth is compounding capital.
Unlike ads that stop when spend stops, well-structured content, user-focused experiences, and earned authority keep working. This is the kind of growth that deserves leadership attention.

• Stop delegating blindly

Imagine this:

You greenlight an SEO strategy with real potential.
You bring in smart people. You approve the roadmap.
You assume it's moving.

But a month goes by, and nothing's changed.
Two months—still stalled.
By the third month, the momentum is gone.

Why?

Because the developer assigned to implement technical fixes is busy building features you requested.
Because the content writer is buried under launch deadlines for a new business line.
Because the designer is heads-down on a new module layout.
Because the video team is producing product content with no time—or knowledge—to embed structured data.

And because SEO was never resourced. It was assigned.

No one stopped to ask:
"Should an SEO lead review this before we ship?"
"Does this follow our search strategy?"
"Will this even be found?"

So the website keeps growing—unoptimized.
Pages are published without schema.
URLs are launched with poor structure.
Headings are out of order.
No internal links, no metadata, no CRO thinking.
Just motion without direction.

And this isn't a one-off. It's a pattern.
The same loop, over and over.
Until someone finally asks, "Why aren't we ranking?"

This is what happens when SEO is treated like a side-task.
When it's handed off with no dedicated hours, no air cover, and no executive follow-up.
When it's *delegated blindly.*

Here's the truth:
If SEO is everyone's job, but no one's priority, it will always come last.

That's not a reflection of your team's ability.
It's a reflection of what you've made possible.

If you've ever assumed it was "being handled"—without checking if the people responsible actually had the time, clarity, or authority to act—then you already know where the problem is.

And now that you see it, you can fix it.

You don't need to write code or optimize content yourself.
You need to lead.
Make SEO visible. Fund it. Assign owners.
Protect its momentum like you would any other growth initiative—because that's what it is.

• Start leading with clarity

If you still think SEO is just about getting found on Google, you're already behind.

The game has changed. Visibility is no longer enough. Today, what happens *after* the click matters just as much as getting it. That's why the future of SEO isn't just about search engines—it's about the **search experience**.

Every click is a test.
Does the page load fast?
Is the headline relevant?
Can the visitor find what they were promised?
Does the layout guide them—or confuse them?
Do the words convert—or just fill space?

That's no longer just SEO as you know it (Search Engine Optimization). That's **Search Experience Optimization (SXO)**—and it can't be owned by one department.

It belongs to everyone.

Developers shape speed and structure.
Writers craft clarity and relevance.
Designers control flow and readability.
Video teams influence time on page.
Marketers align timing and messaging.
Executives decide what gets prioritized—and what gets ignored.

And when all of these things happen in isolation, SEO breaks quietly.
You don't lose rankings overnight—you lose momentum piece by piece.

So no, SEO is no longer a task to be assigned.
It's a standard to be shared.
And the only way to enforce that standard across teams is **through clarity at the top**.

Your job isn't to micromanage.
It's to connect the dots, to define what good looks like, to make SEO visible, not invisible, and to turn every team into a contributor—not an obstacle.

Because if you don't lead with clarity, someone will always ask:
"Wait, didn't we hire someone to handle SEO?"
And that question will keep costing you.

Final Word

You've now seen what most don't: that SEO isn't a line item in marketing—it's a reflection of how your business operates across teams. You've seen where momentum dies, where communication fails, and where leadership—often unintentionally—becomes the bottleneck. But now you also have the power to lead differently.

Because the companies that win in organic search don't just "do SEO."
They commit to it.
They resource it.
They lead it.

So ask yourself:
What are we building right now that no one will ever find?
And what are you going to do about it?

Coming Soon in the SEO Trilogy

This book is only the beginning of a larger project: a trilogy designed to help leaders and teams understand, direct, and scale SEO as a true business asset. The next volumes go deeper into the executive decisions that can either drive or block organic growth, and into the advanced frameworks required to compete in an era dominated by intelligent search engines and AI.

Instead of giving you dates that will quickly become outdated, I invite you to visit SEOtrilogy.com, where you'll always find up-to-date information about the trilogy, including:

- Publication dates for the upcoming books.

- Available formats (print, digital, audiobook, as applicable).

- Where you can purchase the different versions of each book.

- A form to subscribe to the email list and receive sneak peeks, updates, and launch discounts.

- Links to follow me on social media and access additional content: articles, case studies, and resources for leaders.

If you work in SEO, marketing, or product, I'm also interested in hearing about your real-world experiences. At SEOtrilogy.com, you'll find a "Share Your SEO Story" section where you can submit a short summary of a success or failure you've been through. Some of these stories may become case studies—through interview and editing—for future volumes or additional materials.

Participation is completely voluntary, and the details of how these testimonials may be used are clearly explained on that same page.

If this first book helped you see SEO more clearly, SEOtrilogy.com is where you'll find the next step: staying informed about upcoming titles, diving deeper into the topics that matter most to you, and, if you choose, contributing your own experience to this ongoing conversation.

About the Author

Cristobal Varela studied Architecture in Hermosillo, Sonora, Mexico—and never stopped building his future. After moving to the United States, the cost of returning to school pushed him straight into the workforce, starting in a phone sales job with no formal training in technology, marketing, or analytics. Years later, he resumed his education in the U.S., studying for a Bachelor of Science in Information Technology (BSIT) while continuing to grow his career in digital marketing.

From there, he learned business the hard way: growing his own businesses from the ground up, leading small teams through uncertainty, and carrying the pressure that comes when every decision affects payroll, clients, and family. Those years of struggle as a small business owner reshaped how he thinks. They trained him to look at problems from multiple angles, anticipate the impact of executive decisions, and spot strategic risks and opportunities others often miss.

Along the way, he mastered disciplines at the core of modern marketing—photography, videography, graphic design, web development, social media marketing, reputation management, analytics, and Search Engine Optimization (SEO). That blend of creative and technical skills allows him to connect brand, content, and data, turning complex SEO conversations into clear, practical paths leaders can act on.

Today, based in Arizona, he works with national brands in fields such as home construction, medical and pharmaceutical organizations, news companies, and small businesses, helping them align business goals with sustainable, measurable organic growth. This book is a direct result of that journey—written for executives who want to make better decisions about SEO and the teams who depend on those decisions. Learn more at **CristobalVarela.com.**

www.ingramcontent.com/pod-product-compliance
Lightning Source LLC
Chambersburg PA
CBHW061250220326
41599CB00028B/5595